Information Technology Essentials Volume 2

The Beginner's Guide to C#

Eric Frick

Published By Eric Frick 2020
Last Update Sep 2021

Copyright

Contents

1 Introduction

Hello and welcome to Volume 2 of Information Technology Essentials. In this series of books, I lay out a number of core concepts that provide you a strong foundation to become a software engineer. To become a well-rounded software engineer you will need to master a number of concepts, some of these I described in Volume 1 and a number of hands-on skills. One of these skills is learning one or more programming languages. In Volume 2 we will look at developing skills in your first programming language.

In this case we will learn the basics of the C# programming language from Microsoft. C# is one of the most popular programming languages in the world and is a great place to get started. We will be using .NET Core for our development environment.

In this book, I assume that you have no prior programming knowledge and will I take you from basics up to writing a complete program.

One of the great things about .NET Core is that it is a cross-platform development environment that allows you to develop on Microsoft Windows, macOS, and Linux. Therefore, the code examples in this book will run on all three platforms.

In this book, I will start with getting you set up with your development environment. I will give you two options that you can use. First, I will show you how to set up a development environment on a Microsoft Windows PC. You will install .NET core and the Visual Studio Code editor. I will explain the basic operation of the Visual Studio Code editor and how you can use it to edit code. Although I demonstrate with this editor, you can use any code editor to make your programming changes such as Notepad or Notepad ++.

The second option I will show you is how to set up your development environment in the Google Cloud Platform (GCP). Google offers a program for first-time users to sign up for a free account with a $300 credit. Using the cloud is a great way to develop code and it is the method I prefer. If you have never used GCP, this is a great way to get started and to learn a little about cloud computing.

Once you use up your $300 credit, you will incur charges for resources, but they will stop your account when you use this credit up. They also have many always free resources that you can use for development as well. One of the great free resources available is that you can store unlimited source code repositories in GCP for free. In other words, you can store all of your code that you develop in this class in a private repository in GPC for no charge.

Once you build your first program in either environment, you will then make a copy of one exercise to the next so that each lesson builds on the previous one. I have included commands for both the Linux environment (GCP) and the Windows environment to copy those files. The .NET Core commands will be the same for both environments.

In the next two chapters I will show you how to set up and build your first program (Hello World) in both environments to get you started. If you don't want to run in the cloud, you can skip this section. After that, all of the examples will concentrate on the code itself and learning the syntax of the C# programming language.

Learning a programming language requires practice. I tell my students that learning to program is a lot like playing a musical instrument. It requires a lot of practice to get good at it. To get the most out of this book, you should work each of the exercises in each chapter. We will start with a straightforward program and add more pieces to it as I introduce more elements of the C# language.

Each chapter will introduce a new concept and is designed to be a reasonably short exercise. I have also included code listings in this book. If you are reading this as an e-book, you can copy and paste the code from the relevant section if you get stuck. In the last chapter of the book, I have included a link to my website where you can download the code as well.

If this is your first attempt at learning a programming language, you will need to be patient as it will take you a while to get a feel for editing code, and how to debug typos you will introduce while entering in the code. Just take your time and enjoy the process.

Free Access to Videos and Source Code

By purchasing this book, you also get free access to the video version of this course, which I host on my website. You can register for this class using the following link:

https://www.destinlearning.com/p/csharp

Use the code AMAZON2020 for the discount.

If you have any problems registering for this class, please contact me at sales@destinlearning.com.

In this class, I have posted some videos that walk through the exercises, and you can also download the source code as well as all of the examples in this book. Also, you can check out my other video classes on my website as well at http://destinlearning.com

I also have a number of programming-related video examples on my YouTube channel at:

https://youtube.com/destinlearning

2 What is a Console Program?

```
c:\programs>HelloWorld
Hello World!

c:\programs>
```

With C# and .NET you can develop a wide variety of software and project types. You can develop web applications, mobile applications, web services, Windows programs, and several other project types as well. One of the kinds of programs you can create is a console program. A console program is a program, which runs in a console window on the Windows operating system. These programs date back to the early days of the MS-DOS operating system, before PCs had windowed operating systems.

Console programs write characters directly to the screen and read input mainly from the keyboard. Even though these programs are an older style, they are still in widespread use today. They are very useful for running background tasks and for running daily routine tasks. Also, due to their simple nature, they are ideal for learning to program. We will be using console programs in this book to learn the basic syntax of the C# language.

If you decide to move on to develop other types of programs, the concepts and syntax that you will learn in this book are directly transferable. The syntax for C# is the same no matter what kind of program you are writing.

Console programs are not unique only to Windows PCs. Almost all operating systems have some command line capability. This is indeed true on the macOS, as well as the Linux operating system. Also, you can write these programs with a wide variety of languages such as C#, C++, Java, and many others. You can also write these programs with various scripting languages such as PowerShell for Windows, Perl, Python, and many others. However, for this book, we will be running our C# programs in Command Prompt windows. I will show the output of the .NET Core programs that are running in the Linux environment. However, the output will be identical if you are running on macOS or Windows.

3 Hello World Windows

In this chapter, I will show you how to get started using the Windows platform with .NET Core and Visual Studio Code. Visual Studio Code is a lightweight code editor, which is much easier to install and use for C# development, especially for beginners. You can still use Visual Studio if you like it and have it installed. But, for this book, I will describe development using Visual Studio Code.

Download and Install .NET Core

The first step in the process is to download and install .NET Core. In your browser, go to: https://dotnet.microsoft.com/download

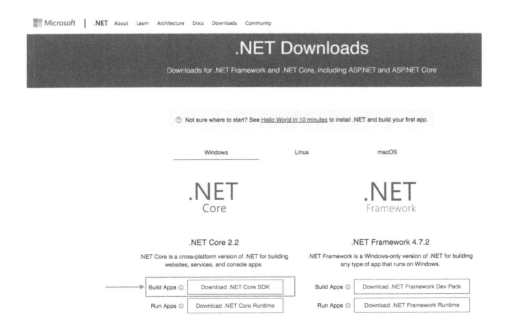

Once you get there, download and install the .NET Core SDK. At the time of the writing of this book, the current release was version 3.1.
Now run the installation file by double-clicking on the file as shown below:

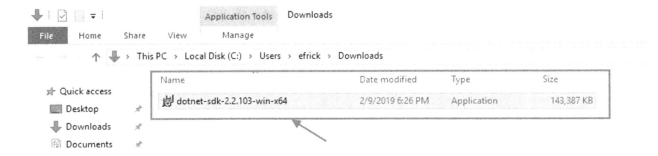

Click the Run button on the Security Warning.

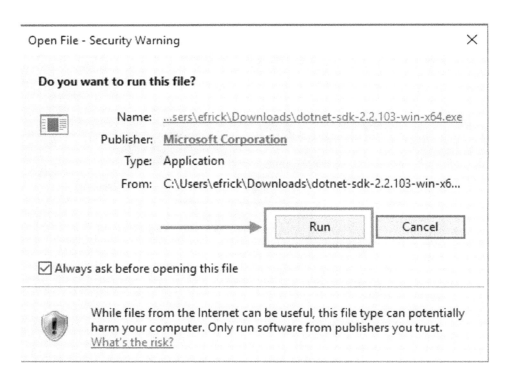

Next, click the Install button as pictured below:

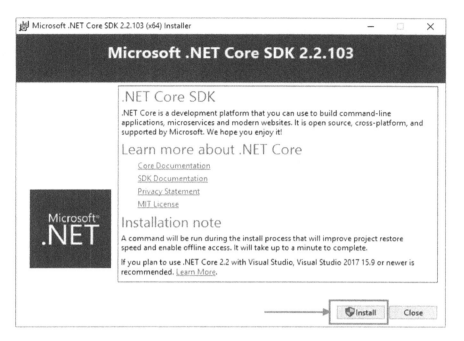

The installer will show you the progress of the installation; it should be just a couple of minutes, depending on the speed of your computer.

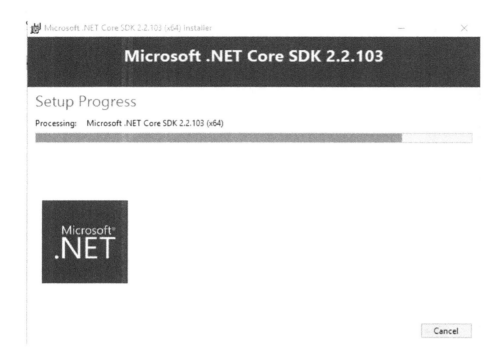

Check to make sure your installation was successful:

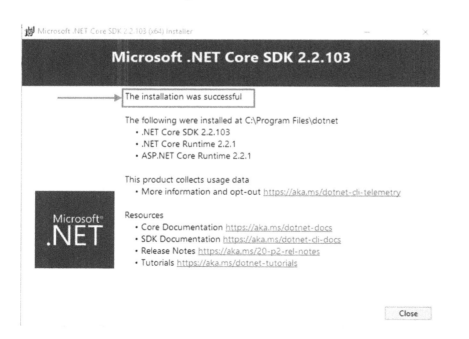

Click the Close button to complete the installation.

Test Your Installation

Open a command prompt window. If you don't know how to do this, you can use the Windows search and find it like in the following example:

Once the command prompt window is open, type in the following command:

```
dotnet --- version
```

We should get the following output:

```
2.2.103
```

Note, your version might be different depending on the time of your installation. Anything 3.1 or later is fine for the examples in this book.

Create a New Project

Now that we are in the command line, in Windows, issue the following command to create a new directory for your projects, and move into that directory.

```
mkdir cscode
cd cscode
```

Now that we have that setup, we can tell .NET Core to create the stub of our first program with the following command.

```
dotnet new console --name ch4
```

We should see the following output after entering this command:

```
Welcome to .NET Core!
---------------------
Learn more about .NET Core: https://aka.ms/dotnet-docs
Use 'dotnet --help' to see available commands or visit:
https://aka.ms/dotnet-cli-docs

Telemetry
---------
The .NET Core tools collect usage data in order to help us improve your
experience. The data is anonymous and doesn't include command-line
arguments. The data is collected by Microsoft and shared with the
community. You can opt-out of telemetry by setting the
DOTNET_CLI_TELEMETRY_OPTOUT environment variable to '1' or 'true' using
your favorite shell.

Read more about .NET Core CLI Tools telemetry:
https://aka.ms/dotnet-cli-telemetry

ASP.NET Core
------------
Successfully installed the ASP.NET Core HTTPS Development Certificate.
To trust the certificate run 'dotnet dev-certs https --trust' (Windows and
macOS only). For establishing trust on other platforms refer to the
platform specific documentation.
For more information on configuring HTTPS see
https://go.microsoft.com/fwlink/?linkid=848054.
Getting ready...
The template "Console Application" was created successfully.

Processing post-creation actions...
Running 'dotnet restore' on ch4\ch4.csproj...
  Restoring packages for C:\Users\efrick\cscode\ch04\ch4\ch4.csproj...
  Generating MSBuild file
C:\Users\efrick\cscode\ch04\cYh4\obj\ch4.csproj.nuget.g.props.
  Generating MSBuild file
C:\Users\efrick\cscode\ch04\ch4\obj\ch4.csproj.nuget.g.targets.
  Restore completed in 367.96 ms for
C:\Users\efrick\cscode\ch04\ch4\ch4.csproj.

Restore succeeded.
```

Open the File in Notepad

We can open the C# file, which was created by the .NET Core command, by first navigating to the correct directory, and then entering the command to edit the file in Notepad.

```
cd ch4
notepad Program.cs
```

Once the file is open in the editor, we should see the following code:

```
Program.cs - Notepad                              —    □    ×
File   Edit   Format   View   Help
using System;

namespace ch4
{
    class Program
    {
        static void Main(string[] args)
        {
            Console.WriteLine("Hello World!");
        }
    }
}
```

Update the Code

Modify the code to add your name at the end of the WriteLine statement, as in the following:

```
using System;

namespace ch4
{
    class Program
    {
        static void Main(string[] args)
        {
            Console.WriteLine("Hello World, my name is Eric!");
        }
    }
}
```

Save and Run Your Program

Click Save under the file menu, as shown below.

On the command line, compile and run your program. First, make sure we are in the correct directory. Issue the cd command with no arguments.

```
cd
```

It should print out the directory like the following:

```
C:\Users\efrick\cscode\ch4
```

If the last directory in the path is ch4, compile and run the program with the following command:

```
dotnet run
```

We should see the following output:

```
Hello World, my name is Eric!
```

Visual Studio Code

Although you can certainly use notepad for all the exercises in this book,Visual Studio Code is a much better editor for this purpose. I will show you the basics of how to install and use Visual Studio Code. First, navigate to the following URL in your browser:
https://code.visualstudio.com/

You should see the following:

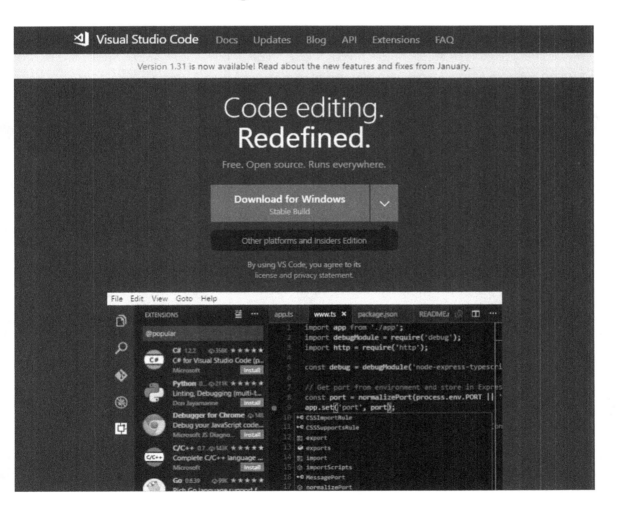

Download the install file for Windows. Once the download has completed, double-click on the downloaded file to launch the installation.

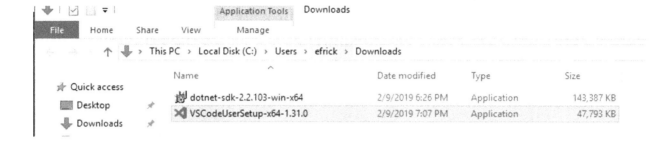

Next, click the Run button on the Security Warning.

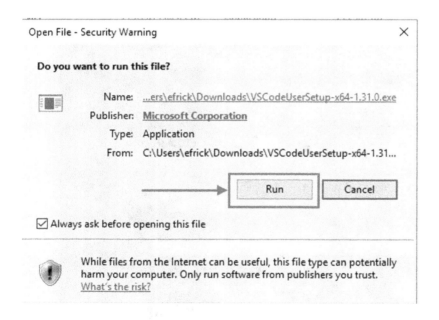

Click the Next button to continue.

Accept the license agreement.

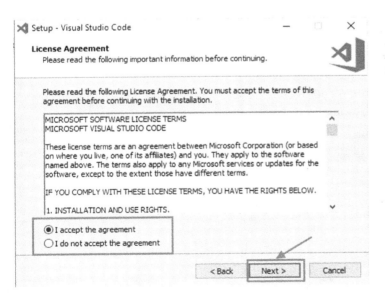

Select the location for the installation. (You can accept the default location)

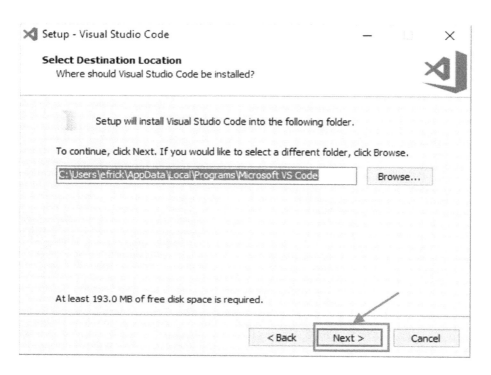

Select the Start menu folder. (You can also accept the default here as well.)

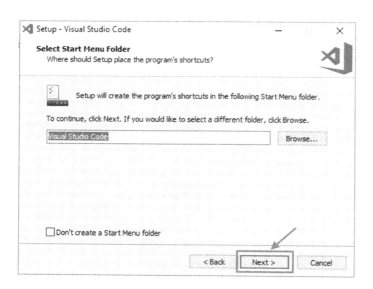

Select any additional options you would like. I selected the 'Create a desktop icon' as well as 'Add to PATH', and click the Next button.

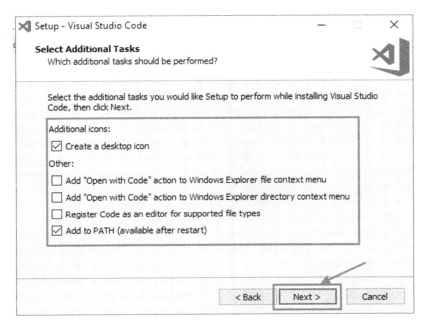

Next, click the Install button.

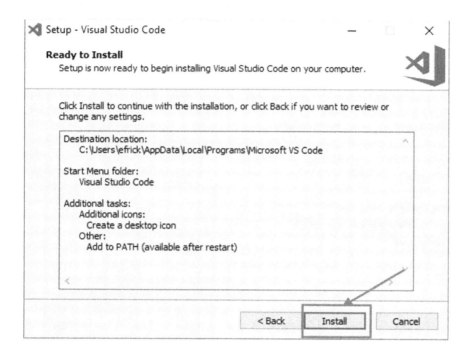

You will now see the following progress bar during the installation:

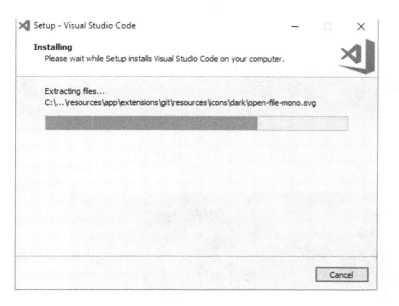

Click Finish to complete the installation.

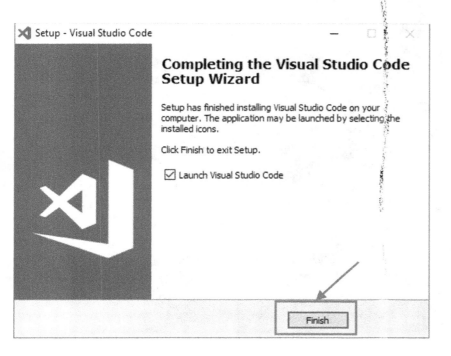

You should now be ready to use Visual Studio Code as an editor. We can test this out by opening a new command window and then navigating to the project we just created.

```
cd \cscode\ch4
code Program.cs
```

We should now see the new code editor with your file loaded:

We now have another option for editing code, instead of using Notepad. Visual Studio Code has many other features than just editing code, but that is beyond the scope of this book. Instead, we will concentrate on the C# language and writing code.

Summary

Congratulations! You have now written your first C# program using Visual Studio Code as your editor! In later chapters, we will expand on this demo, step by step, to add more functions and demonstrate the syntax and capability of the C# programming language.

4 Hello World - Google Cloud Platform

In this chapter, I will show you how to sign up for your free GCP account and write your first program--Hello World. Although you will be running your code in the cloud, your C# code will be exactly the same as the code you would run on your local PC. First, I will show you how to sign up for your GCP account. Following that, I will show you how to set up your code, and compile your first program using the GCP console and cloud editor.

The GCP Cloud Shell is a Linux virtual machine and therefore uses the Linux command shell. I will show you the basic commands to create and copy files in the Linux command shell, but I have not included detailed information on the command shell itself. If you want a basic tutorial on how to use the command shell, I recommend you look at the following tutorial:

https://maker.pro/linux/tutorial/basic-linux-commands-for-beginners

Sign Up for Your GCP Account

First navigate to https://cloud.google.com/free

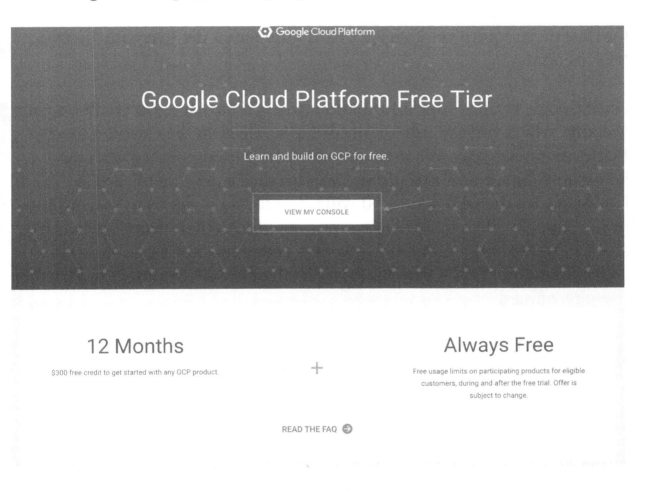

Once you are here, click on the TRY IT FREE button as shown above.

After you have clicked this button, you will see a screen prompt that asks you if you would like to receive more marketing information from Google, and if you agree to the terms and conditions.

After you agree to the terms and conditions, you will enter your name and address, and notate if it is a personal or business account. If you already have a Google account, some of these fields will already be filled in. The last step is to enter a credit card number. Although you will not be charged for this, they require you to have a card number on file to create the free trial account.

Once Google verifies your information, you will have a $300 credit on your account, which you can use for any of the services on the Google Cloud Platform for one year. When the $300 credit is used up, they will stop access to your account and prompt you to enter a new credit card to continue with the paid service. It will be up to you if you want to continue or not. Google offers many, always-free services to their users, such as free storage (5 GB) and unlimited free source code repositories, to store your code and more. I find it very useful to do all of my development work using these resources.

If you would like to read more about the GCP free services, you can follow this link:

https://cloud.google.com/free/docs/gcp-free-tier

The coding methods I demonstrate in this book use the Cloud Console, and will not incur any charges. Using the Cloud Console is free with your account, and the .NET core software is already installed and ready to use. You should now be ready to go with your new Google Cloud account.

Log into Google Cloud Platform

First, navigate to the Google Cloud Platform using the URL
https://cloud.google.com

Log in with your Google account. Once you are logged in, you should see a greeting window with some recent information about GPC service. From there, click on the GO TO CONSOLE button.

You should now be on the console page that shows you a summary of your account.

Open the Cloud Shell

Now that you are logged in to GCP and in the console page, open the Cloud Shell by clicking on the activate Cloud Shell icon in the upper right hand corner of the screen, as illustrated in the following screenshot:

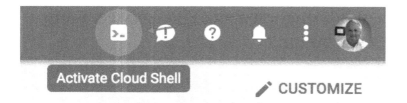

You should now see a screen, as shown below, with the Cloud Shell open at the bottom of the window:

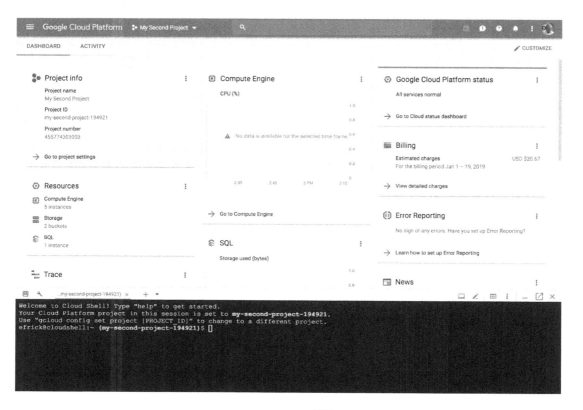

Getting Started

Once you are connected to the Google Cloud Shell, you can verify what version of .NET Core is installed with the following command:

```
dotnet --version
```

It does not matter which version is being displayed, the code in this book will run on any version.

Create a New Project

Now that you are in the command line of Google Cloud Shell, create a new directory for your projects and move into that new directory. Issue the following commands:

```
mkdir cscode
cd ./cscode
```

Now you can use .NET Core to create the stub of your first program with the following command:

```
dotnet new console --name ch4
```

You should see the following output from this command:

```
The template "Console Application" was created successfully.
Processing post-creation actions...
Running 'dotnet restore' on ch4/ch4.csproj...
  Restoring packages for /home/efrick/cscode/ch4/ch4.csproj...
  Generating MSBuild file
/home/efrick/cscode/ch4/obj/ch4.csproj.nuget.g.props.
  Generating MSBuild file
/home/efrick/cscode/ch4/obj/ch4.csproj.nuget.g.targets.
  Restore completed in 207.61 ms for
/home/efrick/cscode/ch4/ch4.csproj.
Restore succeeded.
```

Launch the Code Editor

Now that your base program has been created, edit your program. First, open the editor window in the Google Cloud Shell. You do this by clicking the button, which is located in the upper right hand corner of the window, as pictured below:

You should now see a screen that looks like this:

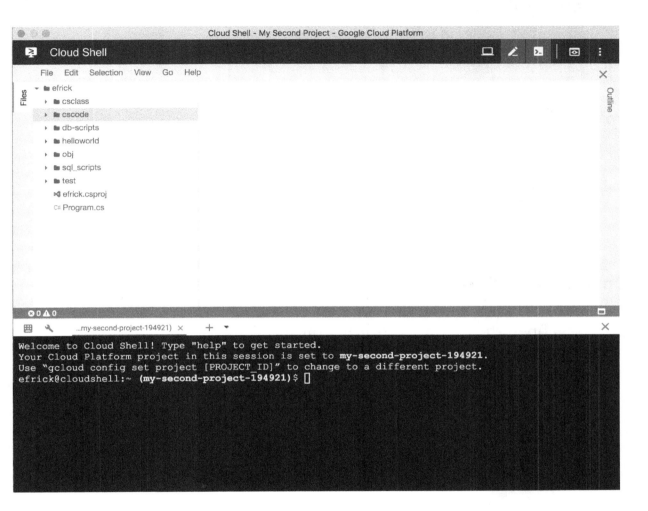

Find Your File

Now that the editor is running, open your file by clicking on the left-hand side of the editor. Choose the file named Program .cs. It will be in a folder called ch4, underneath the cscode folder in your home directory. Reference the screenshot below:

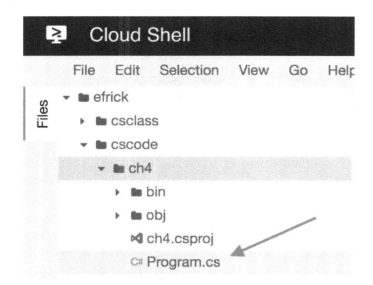

Open in the Editor

Once you open the file in the editor, you should see the following code:

```
File  Edit  Selection  View  Go  Help
▼ ■ efrick                          C# Program.cs ×
  ▶ ■ csclass                        1    using System;
  ▼ ■ cscode                         2
    ▼ ■ ch4                          3    namespace ch4
      ▶ ■ bin                        4    {
      ▶ ■ obj                        5        class Program
        ◁ ch4.csproj                 6        {
        C# Program.cs                7            static void Main(string[] args)
    ▶ ■ ch6                          8            {
    ▶ ■ ch7                          9                Console.WriteLine("Hello World!");
  ▶ ■ db-scripts                    10            }
  ▶ ■ helloworld                    11        }
  ▶ ■ obj                           12    }
  ▶ ■ sql_scripts                   13
  ▶ ■ test
    ◁ efrick.csproj
    C# Program.cs
⊗ 0 ⚠ 0                                          Ln 9, Col 44    Spaces: 4    CSharp
```

Update the Code

Modify the code to add your name at the end of the WriteLine statement, as in the following:

```
using System;

namespace ch4
{
    class Program
    {
        static void Main(string[] args)
        {
            Console.WriteLine("Hello World, my name is Eric!");
        }
    }
}
```

Save and Run Your Program

Click "Save" under the File menu, as shown below:

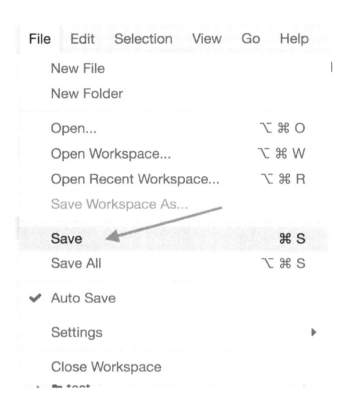

Now, in the command line, compile and run your program. First, make sure you are in the correct directory. Issue the following command:

```
cd ch4
```

Compile and run the program with the following command:

```
dotnet run
```

You should see the following output:

```
Hello World, my name is Eric!
```

Summary

Congratulations! You have now written your first C# program using the GCP console and cloud editor! In later chapters, we will expand on this demo, step by step, to add more functions and demonstrate the syntax and capability of the C# programming language.

5 Comments in C#

```
// Document Your C#
// Code with comments
```

How to Document Your Programs

Comment statements in C# allow programmers to make notes in the code, and document their work. Comments are beneficial in a program when you have to go back and fix a program or add new features long after you have written the program. These comments also help you, and other people, understand the code more quickly and document any unusual features that other developers have added. Comments are also useful for recording and documenting bug fixes.

There are three styles of comments that are available in C#. These are: multi-line comments, single line comments, and XML comments.

Multi-Line Comments

Multi-line comments are used to comment more than one line of code at a time. They can be useful for turning off an entire section of code without deleting the code. The following example (labeled 5.1) is of a multi line comment:

```
using System;

/* 5.1 This is my first program and my first comments
   as well.
   Written by: Eric Frick */

namespace ch4
```

XML Comments

XML comments are a special type of comment in C# and are used to create the documentation of C# code. XML elements (tags) are added into the XML Documentation Comments of C#. The following is an example of an XML comment in a C# program.

```
using System;

/// <summary>
/// This class is my first C# program and is used as an
example.
/// </summary>

namespace ch4
```

You can get more information about XML comments here from Microsoft:

https://docs.microsoft.com/en-us/dotnet/csharp/codedoc

Single Line Comments

Single line comments are perhaps the most common type of comment included in C# code. The syntax is straightforward and illustrated in comment 5.3, in the code below. A single line comment is any line of code, which starts with //. We will be using these types of comments to point out modifications we are making to the code in this book.

You should make a habit of including as many comments as you can, to keep your code well documented. Although it takes a bit of extra time while developing your code, it will pay dividends for you, in the future, when you need to make changes. It will also greatly help in a team environment.

```
using System;

/* 5.1 This is my first program and my first comments as well.
   Written by: Eric Frick */

namespace ch4
{
    class Program
    {
        static void Main(string[] args)
        {
            // 5.3 This is an example of a single line comment
            System.Console.WriteLine("Hello World My Name is
Eric");
        }
    }
}
```

6 Input Statements

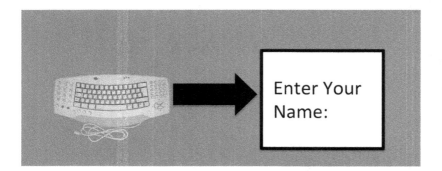

C# provides statements that read input from the user. In a console program, you will use the Console.Readline() statement to read input from the keyboard. The Console.Readline() function returns a string, which you can then use to process in your program. This is best shown by an example.

Start Your Code Editor

We will now make modifications to the project we began in the last chapter. If you are using Windows, open your previous project in either Notepad or Visual Studio Code, as described in Chapter 3. If you are using the Google Cloud Platform (GCP), open that editor as described in Chapter 4.

Open the Hello World Project

Now open the hello world project that we did in the last chapter. Once you have opened our project, you should see the following in your editor:

```
using System;

namespace HelloWorld
{
    class Program
    {
        static void Main(string[] args)
        {
            System.Console.WriteLine("Hello World My Name is Eric");
        }
    }
}
```

Add the Code to Read Your Name

We will now add three new lines of code. I have added comments for each of the changes we will make to the program. Once we have made the changes our program, it should look like this:

```
class Program
    {
        static void Main(string[] args)
        {
            // 6.1 add a string variable for your name
            String myName;

            // 6.2 prompt to enter your name
            System.Console.WriteLine("Please Enter Your Name:");

            // 6.3 now read in your name
            myName = Console.ReadLine();

            // 6.4 add + myName to print out your name
            System.Console.WriteLine("Hello World My Name is:"+myName);
        }
    }
}
```

6.1 For the first line we are adding, we are declaring a string variable, which will be the holding place to store our name.

6.2 Next, we are adding a WriteLine statement to prompt us to enter your name.

6.3 The Console.Readline statement will then read what we type from the keyboard into the myName variable.

6.4 The last modification is to add the + myName in the WriteLine statement. This will actually print out what you entered for your name.

Now Run Your Program

First, save your file in the editor you are using. Then run the program with the following command:

```
dotnet run
```

Verify Your Program Output

You should see the following output in the command prompt window. Note, you will be prompted to type your name and then press the return key, to tell the program you are finished entering data. The program will then print out your name and prompt you to press any key to continue.

```
Please Enter Your Name:
Eric
Hello World My Name is Eric
```

Summary

In this exercise, you learned how to use the Console.ReadLine statement to read data into a string variable, and then print out the contents of that variable.

Link to System.Console.Readline Documentation

I have included a link to the C# reference page about the Console.ReadLine() function.

https://msdn.microsoft.com/en-us/library/system.console.readline%28v=vs.110%29.asp

7 Methods

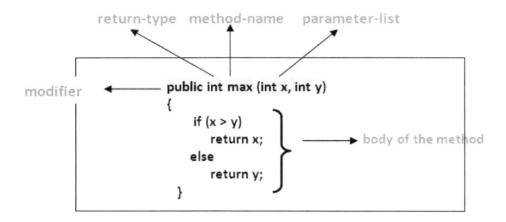

One of the ways to better organize your code is through the use of methods. Methods allow you to break down your programs into much smaller and more manageable pieces. The above code, although very simple, is an example of a method. Our example program, HelloWorld, already has one method--the Main method. Every console application is required to have one method called Main and is the entry point for the program to begin execution.

In this exercise, we will reorganize our code to do exactly the same function as before. But now we will move the prompt to enter our name, and the input statement, into their own method. This method will be called from our Main method.

Syntax

To declare a method in C#, we will use the following syntax:

```
<Access Specifier> <Return Type> <Method Name>(Parameter List)
{
  Method Body
}
```

Here are the definitions of each of these elements:

1. The **Access Specifier** determines the visibility of the method. We will talk more about this in the chapter on object-oriented programming.
2. The **Return Type** determines what, if any, value is returned by the method. In the example above, the method returns an integer.
3. The **Method Name** is simply what the method is called. You will usually name the method something that defines the type or the functionality it is providing.
4. The **Parameter List** (also known as the argument list) is optional, but is a way to pass information into a method from the main program.
5. The last part is the **Method Body**, which is simply the code that comprises the method logic.

Example

We will now simplify our program by adding a method called getName . This will make it easier for us to add more code into our demo program over the next few chapters. First, we need to make a copy of our program, so we can save each of our examples in a separate solution.

Copy the Solution

Navigate to the directory where your source code is stored, and issue the following commands:

Linux (GCP)
```
cp -r ./ch4 ./ch7
cd ./ch7
```

Windows
```
xcopy ch4 ch7
cd ch7
```

Open the code in your editor. You should see the following code:

```
using System;

namespace HelloWorld
{
    class Program
    {
        static void Main(string[] args)
        {
            // 1 add a string variable for your name
            String myName;

            // 2 add a print statement to tell you to enter your
name
            System.Console.WriteLine("Please Enter Your Name:");

            // 3 now the readline statement will read in your name
            myName = Console.ReadLine();

            // 4 add + myName to print out your name
            System.Console.WriteLine("Hello World My Name
is:"+myName);
        }
    }
}
```

Create Your Method

Now we will create our first method. We will make a new method called
GetName, right after the last } of the Main method. Then move the
Readline and WriteLine statement from the Main method into our new
method. We have also modified //1 to call our new method, which allows
the user to enter their name.

```csharp
using System;

namespace HelloWorld
{
    class Program
    {
        static void Main(string[] args)
        {
            // 1 add a string variable for your name
            String myName = GetName();

            // 4 add + myName to print out your name
            System.Console.WriteLine("Hello World My Name is:
"+myName);
        }

        public static string GetName()
        {
            // 1 add a string variable for your name
            String myName;

            // 2 add a print statement to tell you to enter your
name
            System.Console.WriteLine("Please Enter Your Name:");

            // 3 now the readline statement will read in your name
            myName = Console.ReadLine();

            return myName;

        }
    }
}
```

Run Your Program

Make sure you are in the correct directory. Now run your program with the following command:

```
dotnet run
```

You should see the following output:

```
Please Enter Your Name:
Eric
Hello World My Name is: Eric
```

Your program should function the same as before, but the code is now using a method. By using methods, we can organize the code in a much better fashion and make the code more modular. We will use methods heavily in the remainder of this book to add more functionality to our demo program.

Summary

In this exercise, we have added a new method to our program. Although the program behaves exactly as before, it is now broken into two methods, Main and GetName. Although this program is still minimal, you will see how much easier it is to work on your code when you can add one method at a time.

Link to Method Documentation

I have also included a link from the Microsoft documentation that further describes methods and their use.

https://docs.microsoft.com/en-us/dotnet/csharp/programming-guide/classes-and-structs/methods

8 If Statements

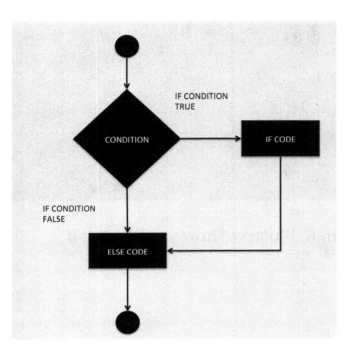

"If" statements in C# are used to implement conditional logic. The C# "if" statement is designed to implement the logic displayed in the picture above. If a particular condition is evaluated to be true, the program will execute a single set of code. If the statement does not evaluate to be true, then multiple sets of codes will be executed. Using these types of statements, programmers can code programs to execute the correct logic to meet the requirements of the program.

Basic Syntax

The basic syntax of an if statement is as follows:

```
if(condition_to_evaluate)
{
   // Code if condition is true
}
else
{
   // Code if condition is false
}
```

We will now add code to our program to illustrate how you can use if statements in a C# program.

Make A Copy of Your Program

Just like in the last chapter, we will now make a copy of our previous program, so we can use this for our next one. We will copy the folder Chapter 7 and rename it Chapter 8. Then we will move into Chapter 8 using the following commands:

Linux (GCP)

```
cp -r ./ch7 ./ch8
cd ./c8
```

Windows

```
copy ch7 ch9
cd ch8
```

Once the file copy command has completed, load the code into your editor. You should now see the following code:

```
Using System;

namespace HelloWorld
{
    class Program
    {
        // Chapter 7
        static void Main(string[] args)
        {
            // 1 add a string variable for your name
            String myName = GetName();

            // 4 add + myName to print out your name
            System.Console.WriteLine("Hello World My Name
is:"+myName);
        }

        public static string GetName()
        {
            // 1 add a string variable for your name
            String myName;

            // 2 add a print statement to tell you to enter your name
            System.Console.WriteLine("Please Enter Your Name:");

            // 3 now the readline statement will read in your name
            myName = Console.ReadLine();
```

```
        return myName;

      }
   }
}
```

Add if Statement

We are now going to add an "if" statement to see if the user enters a blank name, when they enter their name. The code below illustrates the basic syntax and how to use an "if" statement. This code will check the length of the string of the name that is in myName. If it is 0, then the program will print out a message telling the user that they did not enter a name. Otherwise, it will print out a message thanking you for entering your name. The Main method of your program now looks like this:

```
static void Main(string[] args)
{
    // 1 add a string variable for your name
    String myName = GetName();

    // 4 add + myName to print out your name
    System.Console.WriteLine("Hello World My Name is: " + myName);

    if (myName.Length == 0 )
    {
        System.Console.WriteLine("You did not enter your name!");
    }
    else
    {
        System.Console.WriteLine("Thanks for entering your name");
    }

}
```

Run the Program

Now, run the program with the following command:

```
dotnet run
```

You should see the following output:

```
(my-second-project-194921)$ dotnet run
Please Enter Your Name:
Eric
Hello World My Name is:Eric
Thank you for entering your name
```

Now, run the program again, and just press the return key when entering your name. You should see the following output:

```
efrick@cloudshell:~/cscode/ch8 (my-second-project-194921)$
dotnet run
Please Enter Your Name:
Hello World My Name is:
You did not enter your name!
```

Summary

You have now coded a basic "if" statement into a C# program, and have seen how to use them for conditional logic. These are some of the most commonly used statements in computer programs.

Link to Method Documentation

I have also included a link to the Microsoft C# documentation on if statements for additional information and examples.

https://docs.microsoft.com/en-us/dotnet/csharp/language-reference/keywords/if-else

9 Loops

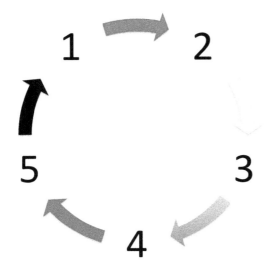

Loops in C# are used to repeat a block of code until a condition is met. They are instrumental in processing many routine programming tasks. For example, it might be helpful to repeat a block of code to process bills or invoices for each day of the month. Another example might be to calculate grades for each student in a class. Loops can also be nested to process multiple collections of data. Using the previous example, you might want to process grades for all courses in a school, and then calculate grades for each student in a class. This type of application is an example of a nested loop.

Basic Syntax

There are three types of loops in C#. The first is a "for" loop.

```
// In this example the WriteLine statement will be executed 10
times
// The int i portion declares a variable of type integer; next
it // will loop while i is less than or equal to 10
// Finally i++ increments I by one each time the loop is
executed
for (int i = 1; i<=10; i++)
{
    System.Console.WriteLine(i);
}
```

The next type of loop is a "while" loop.

```
// In this loop while the condition is true the code inside the
brackets will continue to be executed
while (condition)
{
    // Statements to be repeated
}
```

The last type of loop is a "do" loop. This code is almost the same as the "while" loop, except the condition is checked at the bottom of the loop instead of the top of the loop.

```
do
{
    // statements
}
while(condition);  // Don't forget the trailing ; in do...while
loops
```

To clarify how to use a loop in a program, we will add a "while" loop to our Main method, to demonstrate its use.

Make a Copy of Your Program

As in the previous chapters, we will start by making a copy of our last program and place it in a new folder called Chapter 9. Use the following commands to make a copy of your Chapter 8 program and move the copy to a new folder for Chapter 9.

Linux (CGP)

```
cp -r ./ch8 ./ch9
cd ./ch9
```

Windows

```
xcopy ch8 ch9
cd ch9
```

Once the file copy has completed, load the Program.cs file into your editor. You should see the following code.

```
static void Main(string[] args)
{
        // 1 add a string variable for your name
        String myName = GetName();

        // 4 add + myName to print out your name
        System.Console.WriteLine("Hello World My Name is: " +
myName);

        if (myName.Length == 0 )
        {
            System.Console.WriteLine("You did not enter your
name!");
        }
        else
        {
            System.Console.WriteLine("Thanks for entering your
name");
        }

}
```

Add Code for Loop

We are now going to add some code to demonstrate the use of a loop. Our loop will continue to ask for your name until you enter the name EXIT. Once you enter this for your name, the program will quit. Here are the lines we will be adding to the Main method:

```
// Chapter 9
Static void Main(string[] args)
{
    // 9.1 control variable for program
    Boolean done = false;

    // 9.2 loop to control flow of program
    while(done==false)
    {

        // 1 add a string variable for your name
        String myName = GetName();

        // 4 add + myName to print out your name
        System.Console.WriteLine("Hello World My Name is: "+myName);

        // 9.3 check for exit condition
        if (myName == "EXIT")
        {
            System.Console.WriteLine("Thanks and goodbye!");
            done = true;
        }
        else
        {
            System.Console.WriteLine("Thanks for entering your name");
        }
```

```
    } // 9.4 end of loop code
}
```

9. 1 First we are declaring a variable called "done" that is a Boolean variable. Boolean variables can only have a value of true and false.

9.2 Next, we are adding the "while" loop that will continue to execute the code until the condition is false.

9.2 This is the beginning bracket of the code we want to repeat.

9.3 When the user enters EXIT, we will kick out of the loop and end the program.

9.4 This is the end bracket of the code we want to repeat. Notice the code is indented so it is easier to read.

Please note, we are removing the "if" logic from the previous chapter. (You can modify the "if else" statements, which are already in place.

Run Your Program

Now run your program using the following command:

```
dotnet run
```

Enter your name and then EXIT to complete the program. You should see the following output:

```
Please Enter Your Name:
Eric
Hello World My Name is: Eric
Thank you for entering your name
Please Enter Your Name:
EXIT
Hello World My Name is: EXIT
Thanks and goodbye!
```

You have now seen how to use a loop in our sample program. Loops are one of the most common constructs used in computer programs. One of the techniques that will help you while coding with loops, is to have your code indented properly, so it is easy to read. In case you have problems entering the code above, I have included a complete listing of the program here so you can copy and paste if necessary, to help fix any typos you have.

```
using System;

namespace HelloWorld
{
    class Program
    {
        // code with Chapter 9 modifications
        static void Main(string[] args)
        {
            // 9.1 control variable for program
            Boolean done = false;

            // 9.2 loop to control flow of program
            while(done==false)
            {
```

```
            // 1 add a string variable for your name
            String myName = GetName();

            // 4 add + myName to print out your name
            System.Console.WriteLine("Hello World My Name is: "+myName);

             // 9.3 check for exit condition
             if (myName == "EXIT")
             {
                 System.Console.WriteLine("Thanks and goodbye!");
                 done = true;
             }
             else
             {
                 System.Console.WriteLine("Thank you for entering your
name");

             }

        } // 9.4 end of loop code
    }

    public static string GetName()
    {
        // 1 add a string variable for your name
        String myName;

        // 2 add a print statement to tell you to enter your name
        System.Console.WriteLine("Please Enter Your Name:"); Name:");

        // 3 now the readline statement will read in your name
        myName = Console.ReadLine();

        Return myName;


    }
  }
}
```

Summary

In this chapter, we have discussed loops and how you can use them in programs. As an example of this, we added a simple "while" loop to control the program flow within the Main method of our demo program.

Link to Loop Documentation

I have included a link to the C# loop documentation from Microsoft for further reading and investigation.

https://msdn.microsoft.com/en-us/library/ms228598(v=vs.90).aspx

Coding Exercise 1

The following are coding exercises to help you reinforce the C# statements we have studied so far.

Exercise 1.1

Write a C# program to swap two numbers.

Input	Expected Output
Input the First Number : 5 Input the Second Number : 6	First Number : 6 Second Number : 5

Exercise 1.2

Write a C# program to print the result of the specified operations.

Input	Expected Output
-1 + 4 * 6 (35+ 5) % 7 14 + -4 * 6 / 11 2 + 15 / 6 * 1 - 7 % 2	23 5 12 3

Exercise 1.3

Write a C# program that will print out the numbers from 20 to 1 in reverse order.

Input	Expected Output
	20,19,18,17,16,15,14,13,12,11,10,9,8, 7,6,5,4,3,2,1

Exercise 1.4

Write a C# program that reads in ten numbers from the user and calculates their sum and average.

Input	Expected Output
1,2,3,4,5,6,7,8,9,10	The sum of the numbers is: 55 The average of the numbers is: 5.5

10 Data Types

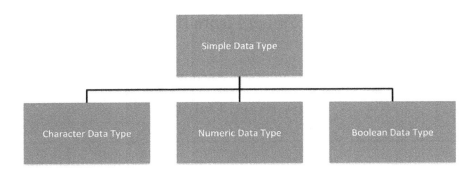

C# defines a wide variety of data types to support a robust programming model. There are many different types of built-in data types to support basic processing. We have already used boolean and string variables in our program, and we will add more as we go further and add to our demo program. In addition to these data types, there are built-in types to handle integer values, date and times, floating-point values for scientific calculations, and many other types. At the end of this chapter, I have included a reference link, so you can see the list of built-in data types from the Microsoft documentation.

Basic Syntax

The basic syntax for declaring a variable in a C# program is as follows:

DataTypeName VariableName;

First, you declare the type of variable you want, and then you assign it a name. In the following example, I am declaring a variable of type int, called iselection.

```
int iselection;
```

You can also initialize the variable at the same time you define it. To do this, you can use the following syntax:

```
int iselection = 0;
```

Please remember the definition must end with a semicolon, and variable names are case-sensitive in C#. ISelection is different from iselection. Many programmers have difficulty getting used to this with C#.

Data Types in C#

The following table lists all the available data types in C# and the range of values they each support:

Type	Represents	Range	Default Value
bool	Boolean value	True or False	False
byte	8-bit unsigned integer	0 to 255	0
char	16-bit Unicode character	U +0000 to U +ffff	'\0'
decimal	128-bit precise decimal values with 28-29 significant digits	$(-7.9 \times 10^{28}$ to $7.9 \times 10^{28}) / 10^{0}$to 28	0.0M
double	64-bit double-precision floating-point type	$(+/-)5.0 \times 10^{-324}$ to $(+/-)1.7 \times 10^{30}$	0.0F
float	32-bit single-precision floating-point type	-3.4×10^{38} to $+ 3.4 \times 10^{38}$	0.0F
int	32-bit signed integer type	-2,147,483,648 to 2,147,483,647	0
long	64-bit signed integer type	-9,223,372,036,854,775,808 to 9,223,372,036,854,775,807	0L
sbyte	8-bit signed integer type	-128 to 127	0
short	16-bit signed integer type	32,768 to 32,767	0
uint	32-bit unsigned integer type	0 to 4,294,967,295	0

ulong	64-bit unsigned integer type	0 to 18,446,744,073,709,551 ,615	0
ushort	16-bit unsigned integer type	0 to 65,535	0

We will first start with a simple example, in this chapter, to illustrate the use of data types. We will make a simple change in our program to introduce the use of the integer data type. Over the next few chapters, we will begin to add more variables, of different types, to illustrate their use in C# programs. So, in the next example, we continue to extend our sample program by adding another new feature to it.

Make a Copy of Your Program

Make a copy of your last program called Chapter 9 and move the copy to Chapter 10, using the following commands:

Linux (GCP)

```
cp -r ./ch9 ./ch10
cd ./ch10
```

Windows

```
xcopy ch9 ch10
cd ch10
```

After doing this, load the program file into your editor, and you should see the following code:

```
// Chapter 9 Main method code
static void Main(string[] args)
{
    Boolean done = false;

    while (done == false)
    {
        // 1 add a string variable for your name
        String myName = GetName();

        // 4 add + myName at the end of the line to print out your
name
        System.Console.WriteLine("Hello World My Name is:" +
myName);

        // check to see if we are done with the program
        if (myName == "EXIT")
        {
            System.Console.WriteLine("Thank you and goodbye!");
            done = true;
        }
        else
        {
            System.Console.WriteLine("Thank you for entering your
name");
        }
    }
}
```

Make Changes to the Main Method

We are now going to add a new variable of type int, indicated by the // 10.1 comment. The new line you are adding is displayed below. This line defines a new variable of type int and initializes it to 0.

```
Static void Main(string[] args)
{
    Boolean done = false; // 10.1
    int iselection = 0;
```

Next, we are going to add some new code to the Main method, so it will display a menu to the users. They can then select what they want to do in this program. Following that, we will adjust the "if" statements, which follow the menu, to allow the program to act correctly, based on the input. Make the following changes to the program:

```
while(done==false)
  {
      // 10.2
      System.Console.WriteLine("1) Select to enter your name");
      System.Console.WriteLine("9) Select to exit the program");
      System.Console.WriteLine("Please Enter Your Selection:");

      // 10.3 read in a string and convert it to an integer
      iselection = Convert.ToInt32(System.Console.ReadLine());

      // 10.4 check for menu selection 1
      if(iselection==1)
      {
          // 1 add a string variable for your name
          String myName = GetName();

          // 4 add + myName to print out your name
          System.Console.WriteLine("Hello World My Name is:
"+myName);
      }

      // 10.5 check for exit condition
      if (iselection == 9)
      {
          System.Console.WriteLine("Thanks and goodbye!");
          done = true;
      }

  } // 10.6 end of loop code
```

Here is a summary of the new code we just added:

10.2 The first series of WriteLine statements print out a menu for the users.

10.3 Next, we are reading data into our new variable--iselection. Please note: We must convert the data to a type of int, since the ReadLine method returns a string.

10.4 Add a new if statement to see if a "1" was selected. If it was, ask the user for their name.

10.5 Modify the if statement to check if "9" was entered by the user. If it was, set done to true and exit the program.

10.6 This is the end of our loop. We no longer need the else statement that was here before.

Run Your Program

Now run your program with the following command after saving your changes:

```
dotnet run
```

Verify that if you enter 1, it asks for your name. Also, verify if you enter 9, the program exits. If you enter other values, what happens? Why? We cover this scenario in more detail in upcoming chapters.

```
efrick@cloudshell:~/cscode/ch10 (my-second-project-194921)$
dotnet run
1) Select to enter your name
9) Select to exit the program
Please Enter Your Selection:
1
Please Enter Your Name:
Eric
Hello World My Name is: Eric
1) Select to enter your name
9) Select to exit the program
Please Enter Your Selection:
9
Thanks and goodbye!
```

Save Your Program

Don't forget to save all of the changes you have made to this program, once you have it running correctly. Here is the complete listing of the Main method of our code, to this point.

```
// Chapter 10
static void Main(string[] args)
{

    // 9.1 control variable for program
    Boolean done = false;
    int iselection = 0;

    // 9.2 add string variable for your name
    while(done==false)
    {
        // 10.1 print out menu
        System.Console.WriteLine("1) Select to enter your name");
        System.Console.WriteLine("9) Select to exit the program");
        System.Console.WriteLine("Please Enter Your Selection:");

        // 10.2 read in a string and convert it to an integer
        iselection = Convert.ToInt32(System.Console.ReadLine());

        // 10.3 check for menu selection 1
        if(iselection==1)
        {
            // 1 add a string variable for your name
            String myName = GetName();

            // 4 add + myName to print out your name
            System.Console.WriteLine("Hello World My Name is:
"+myName);
        }

        // 9.3 check for exit condition
        if (iselection == 9)
        {
```

```
            System.Console.WriteLine("Thanks and goodbye!");
            done = true;
        }

    } // 10.4 end of loop code
}
```

Link to Loop Documentation

I have included a link to the C# documentation on data types from Microsoft, for further reading.

https://docs.microsoft.com/en-us/dotnet/csharp/language-reference/keywords/built-in-types-table

11 Exceptions

C# and .NET offer a robust way of handling runtime errors, which you can encounter in a program. Nothing is more annoying to users than a program that crashes while it is running. Exceptions are a way to gracefully handle any unforeseen errors that can occur in a program.

Basic Syntax

The code below illustrates the basic syntax for an exception handler. It is pretty simple and straightforward.

```
try
{
   // Some code to execute
}
catch
{
   // Code to process the error condition
}
```

This code shows the basic syntax. There are other options for this code, but for our purposes, we will demonstrate the basic capabilities of exception handling, and how to code an exception handler in this chapter.

Copy Your Program

First, make a copy of the program from the previous chapter--Chapter 10.
Move the copy to its new folder--Chapter 11. Use the following commands:

Linux (GCP)

```
cp -r ./ch10 ./ch11
cd ./ch10
```

Windows

```
xcopy ch10 ch1
cd ch11
```

After opening your editor, you should see the following code:

```
static void Main(string[] args)
  {
      Boolean done = false;
      int iselection = 0;


      while (done == false)
      {
          System.Console.WriteLine("1) Select to enter your name");
          System.Console.WriteLine("9) Select to exit to program");
          System.Console.WriteLine("Please Enter Your Selection:");

          // read in a string and convert it to an integer
          iselection = Convert.ToInt32(System.Console.ReadLine());
```

Demonstrate an Exception

Now that your program has been loaded into the editor, go ahead and run the program by issuing the following command:

```
dotnet run
```

You should see the following output in our console program, asking us for our selection:

```
efrick@cloudshell:~/cscode/ch11 (my-second-project-194921)$
dotnet run
1) Select to enter your name
9) Select to exit the program
Please Enter Your Selection:
```

Instead of entering 1 or 2 as a valid input, enter the letter A and press return. You should now get an error message, just like the following output:

```
efrick@cloudshell:~/cscode/ch11 (my-second-project-194921)$
dotnet run
1) Select to enter your name
9) Select to exit the program
Please Enter Your Selection:
A
Unhandled Exception: System.FormatException: Input string was
not in a correct format.
   at System.Number.StringToNumber(ReadOnlySpan`1 str,
NumberStyles options, NumberBuffer& number, NumberFormatInfo
info, Boolean parseDecimal)
   at System.Number.ParseInt32(ReadOnlySpan`1 s, NumberStyles
style, NumberFormatInfo info)
   at System.Convert.ToInt32(String value)
   at HelloWorld.Program.Main(String[] args) in
/home/efrick/cscode/ch11/Program.cs:line 24
```

What happened? We tried to convert a character number into an integer, which is an illegal operation in C#, and caused an error.

Code the Fix

We can now code an exception handler to handle the problem gracefully.
Enter some new code, as shown in the following example:

```
// 11.1 validate the integer conversion process
try
{
    // 10.2 read in a string and convert it to an integer
    iselection = Convert.ToInt32(System.Console.ReadLine());
}
// 11.2 catch the conversion error
catch
{
    // 11.3 notify the user of the error
    System.Console.WriteLine("You have entered an invalid
selection");
}
```

Add the code as follows:

11.1 Add the try along with the braces, to isolate the conversion to the
integer and ReadLine process.

11.2 Add the catch portion that executes if an error occurs.

11.3 A WriteLine statement notifies the user that an error has occurred,
and the system loops, asking again for a menu selection.

Run the New Code

If we now run our new code, we can see if the code is working correctly, by catching the previous error and allowing the user to make another selection. You should see output like the following in your program:

```
efrick@cloudshell:~/cscode/ch11 (my-second-project-194921)$
dotnet run
1) Select to enter your name
9) Select to exit the program
Please Enter Your Selection:
A
You have entered an invalid selection
1) Select to enter your name
9) Select to exit the program
Please Enter Your Selection:
```

Save Your Program

Don't forget to save all of the changes you have made to this program, once you have it running correctly.

Summary

Listed below is the complete code for the Main method. If you have trouble entering your new code, you can use this to copy and paste pieces of it to help get you going.

```csharp
// Chapter 11
static void Main(string[] args)
{
    Boolean done = false;
    int iselection = 0;


    while (done == false)
    {
        System.Console.WriteLine("1) Select to enter in your
name");
        System.Console.WriteLine("9) Select to exit to
program");
        System.Console.WriteLine("Please Enter Your
Selection:");

        // read in a string and convert it to an integer now an
validate it can be converted
        try
        {
            iselection =
Convert.ToInt32(System.Console.ReadLine());
        }
        catch
        {
            System.Console.WriteLine("You have entered an
invalid selection.");
        }
        if (iselection == 1)
        {
            // 1 add a string variable for your name
            String myName = GetName();

            // 4 add + myName at the end of the line to print
```

```
 out your name
            System.Console.WriteLine("Hello World My Name is:" +
myName);
        }

        // check to see if we are done with the program
        // change this to use an integer now
        if (iselection == 9)
        {
            System.Console.WriteLine("Thank you and goodbye!");
            done = true;
        }

    }

}
```

Link to Exception Documentation

I have included a link to the C# documentation on exception handling from Microsoft, for further reading.

https://msdn.microsoft.com/en-us/library/ms229005(v=vs.100).aspx

12 Switch Statement

C# includes a switch statement, which allows a selection to execute against a list of candidates. A switch statement offers a simplified syntax, which makes it easier to read when a single condition applies to a series of alternatives. Many other programming languages also include switch statements.

Basic Syntax

The following code illustrates the basic syntax for a switch statement:

```
int caseSwitch = 1;

    switch (caseSwitch)
    {
        case 1:
            Console.WriteLine("Case 1");
            break;
        case 2:
            Console.WriteLine("Case 2");
            break;
        default:
            Console.WriteLine("Default case");
            break;
    }
```

This code shows the basic syntax for the switch statement. To illustrate this further, and to simplify the code, we will now add this capability into our

Main method. We will add more menu selections in later chapters of this book, so this is a good time to simplify our code.

Copy Your Program

First make a copy of the program from the previous chapter, Chapter 11, and move the copy to a new folder called Chapter 12. Use the following commands:

Linux (GCP)

```
cp -r ./ch11 ./ch12
cd ./ch12
```

Windows

```
xcopy ch11 ch12
cd ch12
```

Once you open your editor, you should see the following code:

```
// Chapter 11
class Program
{
    static void Main(string[] args)
    {

        // 9.1 control variable for program
        Boolean done = false;
        int iselection = 0;

        // 9.2 add string variable for your name
        while(done==false)
        {
            // 10.1 print out menu
            System.Console.WriteLine("1) Select to enter your
name");
            System.Console.WriteLine("9) Select to exit the
program");
            System.Console.WriteLine("Please Enter Your
Selection:");

            // 11.1 validate the integer conversion process
            try
            {
                // 10.2 read in a string and convert it to an
integer
                iselection =
Convert.ToInt32(System.Console.ReadLine());
            }
            // 11.2 catch the conversion error
            catch
            {
                // 11.3 notify the user of the error
```

```
            System.Console.WriteLine("You have entered an
invalid selection");
        }

        // 10.3 check for menu selection 1
        if(iselection==1)
        {
          // 1 add a string variable for your name
          String myName = GetName();

          // 4 add + myName to print out your name
          System.Console.WriteLine("Hello World My Name is:
"+myName);

        }

        // 9.3 check for exit condition
        if (iselection == 9)
        {
            System.Console.WriteLine("Thanks and goodbye!");
            done = true;
        }

    } // 10.4 end of loop code

  }
```

We will now add our new code right underneath comment 10.3, in the above listing. We will replace the series of "if" statements with our new switch statement.

Our new code, with the switch statement, should look like the following listing:

```
// 10.3 check for menu selection 1
switch (iselection) // 12.1
{
    case 1: // 12.2
    {
        // 1 add a string variable for your name
        String myName = GetName();

        // 4 add + myName to print out your name
        System.Console.WriteLine("Hello World My Name is:" +
myName);
        break;
    }

    // check to see if we are done with the program
    // change this to use an integer now
    case 9: // 12.3
    {
        System.Console.WriteLine("Thank you and goodbye!");
        done = true;
        break;
    }

    default: // 12.4 handle the default case
    {
        System.Console.WriteLine("You have made an invalid
selection.");
        break;
    }
} // 12.5
```

The new code is a bit more organized, and is based on the following:

12.1 The switch statement includes all of the choices for the menu.

12.2 Each selection is represented with a case statement enclosed by { }.

12.3 Make sure to include a case for each selection.

12.4 The default case is matched if none of the previous cases match. In this case, it prints out an error message.

12.5 Make sure a beginning { and an ending } encloses the case statement.

Run Your Program

Now, run your program with the following command:

```
dotnet run
```

You should have the same functionality as the program from the previous chapter, but running the new case statement. If you are having any problems, the complete listing for this program is in the next section.

Code Listing

In case you have problems entering the code above, I have included a complete listing of the Main methods here, so you can copy and paste if necessary, to help fix any typos you have.

```
// Chapter 12
static void Main(string[] args)
    {

        // 9.1 control variable for program
        Boolean done = false;
        int iselection = 0;

        // 9.2 add string variable for your name
        while(done==false)
        {
            // 10.1 print out menu
            System.Console.WriteLine("1) Select to enter your
name");
            System.Console.WriteLine("9) Select to exit the
program");
            System.Console.WriteLine("Please Enter Your
Selection:");

            // 11.1 validate the integer conversion process
            try
            {
                // 10.2 read in a string and convert it to an
integer
                iselection =
Convert.ToInt32(System.Console.ReadLine());
            }
            // 11.2 catch the conversion error
            catch
            {
```

```
                // 11.3 notify the user of the error
                System.Console.WriteLine("You have entered an
invalid selection");
            }

            // 10.3 check for menu selection 1
            switch (iselection) // 12.1
            {
                case 1: // 12.2
                {
                    // 1 add a string variable for your name
                    String myName = GetName();
                    // 4 add + myName to print out your name
                    System.Console.WriteLine("Hello World My Name
is:" + myName);

                    break;
                }
                // check to see if we are done with the program
                // change this to use an integer now
                case 9: // 12.3
                {
                    System.Console.WriteLine("Thank you and
goodbye!");

                    done = true;
                    break;
                }
                default: // 12.4 handle the default case
                {
                    System.Console.WriteLine("You have made an
invalid selection.");
```

Save Your Program

Once you get your program running, don't forget to save all of your
changes in your code editor.

Summary

You have now seen the basic operation of the switch statement in C#. If you have many choices for the same selection, using this technique, instead of a series of "if" statements, results in code, which is much easier to read and understand.

Link to Switch Documentation

For more information on switch statements, I have included a link to the C# documentation from the C# reference.

https://docs.microsoft.com/en-us/dotnet/csharp/language-reference/keywords/switch

13 Arrays

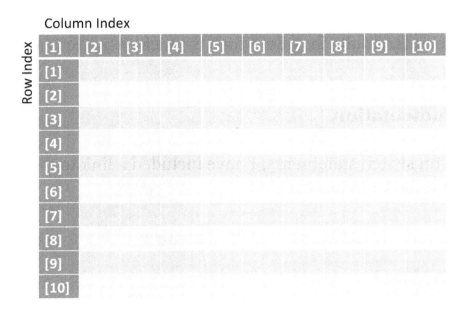

In C#, you can store multiple variables of the same type, in a collection called an array. You declare an array by declaring the type of the array and the number of elements. You can also specify the values in the array when you initially declare it, as well. A single-dimensional array is just a column of values. A two dimensional array is similar to a spreadsheet with rows and columns. A three-dimensional array would be similar to having rows, columns, and tabs on a spreadsheet. You can expand an array to beyond three dimensions, but at some point, they become harder to manage, when they get very large. In this chapter, I will illustrate an example with a single-dimensional array.

Basic Syntax

The basic syntax for an array declaration is as follows:

```
type[] arrayName;
```

In the following example, I am declaring an integer array with five values.

```
int[] scores = new int[] {95, 92, 88, 81, 75};
```

In the next example, I am declaring a string array with seven values.

```
String weekdays = new string[] {"Sun", "Mon", "Tue", "Wed",
"Thr", "Fri", "Sat"};
```

Copy Your Program

First make a copy of the program from the previous chapter, Chapter 12, and move the copy to a new folder, Chapter 13. Use the following commands:

Linux (GCP)

```
cp -r ./ch12 ./ch13
cd ./ch13
```

Windows

```
xcopy ch12 ch13
cd ch13
```

Once you have loaded your program into your editor, you should see the following code:

```
static void Main(string[] args)
{

    // 9.1 control variable for program
    Boolean done = false;
    int iselection = 0;

    // 9.2 add string variable for your name
    while(done==false)
    {
        // 10.1 print out menu
        System.Console.WriteLine("1) Select to enter your name");
        System.Console.WriteLine("9) Select to exit the program");
        System.Console.WriteLine("Please Enter Your Selection:");
```

We are going to insert a new print statement into our menu, adding a new option. Here is the new line we are adding:

```
// 9.2 add string variable for your name
while(done==false)
{
    // 10.1 print out menu
    System.Console.WriteLine("1) Select to enter your name");
    System.Console.WriteLine("2) Array demo");
    System.Console.WriteLine("9) Select to exit the program");
    System.Console.WriteLine("Please Enter Your Selection:");
```

Now that we have added this line, we need to add some new code to handle our new selection:

```
// this will demonstrate the capability of arrays
case 2:              // 13.3
{
    ArrayDemo();   // 13.2
    break;         // 13.3
}
```

Above is the new code we added for our new case.

13.1 Add the new case for the menu item.

13.2 Add a call to our new method called ArrayDemo(). (Note, we have not written the code for this method yet.)

13.3 Don't forget the break statement.

Add Our New Method

Now, let's insert code for our new method, right after the Main method, but before the GetName method, as in the following figure:

```
        } // 10.4 end of loop code
    }

    // insert new code here

    public static string GetName()
    {
```

Now we can add the code for our new method:

```
// show the basic operation of an array
public static void ArrayDemo()
{
  // 13.2 set up a string for each day of the week
  string[] days = new string[]
{"Sun","Mon","Tue","Wed","Thr","Fri","Sat"};

  int i;

  // 13.3 print out all of the days
  for(i=0;i<=6;i++)
  {
    System.Console.WriteLine("The day of the week is: " + days[i]);
  }

}
```

Notice the following about our new code:

13.2 This is the standard declaration for a string array. Specify the values along with the declaration.

13.3 Print out all of the values in the array note, which in C# arrays are 0 based. That is, the first element in the array is element zero, and therefore the last element in the array is the size -1. (In our case that last day is element 6). The for loop, loops through each element in the array and prints it out to the screen.

Run Your Program

Now run the program to make sure it is working correctly.

```
dotnet run
```

If all goes well you should get the following output:

```
efrick@cloudshell:~/cscode/ch13 (my-second-project-194921)$
dotnet run
1) Select to enter your name
2) Array demo
9) Select to exit the program
Please Enter Your Selection:
2

The day of the week is: Sun
The day of the week is: Mon
The day of the week is: Tue
The day of the week is: Wed
```

```
The day of the week is: Thr
The day of the week is: Fri
The day of the week is: Sat
1) Select to enter your name
2) Array demo
9) Select to exit the program
Please Enter Your Selection:
9
Thank you and goodbye!
```

Save Your Program

Once you get your program running, make sure you save all of the changes in your editor.

Summary

You have now seen how a basic, one-dimensional array functions in C#. You can also have multidimensional arrays, as well as arrays of objects. We will see more arrays later in this book. This code, however, demonstrates the basics of array processing in C#. The following is a complete listing of the entire program for this exercise. You can copy and paste portions into your program if you get stuck.

```csharp
using System;

namespace HelloWorld
{
    // Chapter 13
    class Program
    {
        static void Main(string[] args)
        {

            // 9.1 control variable for program
            Boolean done = false;
            int iselection = 0;

            // 9.2 add string variable for your name
            while(done==false)
            {
                // 10.1 print out menu
                System.Console.WriteLine("1) Select to enter your
name");
                System.Console.WriteLine("2) Array demo");
                System.Console.WriteLine("9) Select to exit the
program");
                System.Console.WriteLine("Please Enter Your
Selection:");

                // 11.1 validate the integer conversion process
                try
                {
                    // 10.2 read in a string and convert it to an
integer
                    iselection =
Convert.ToInt32(System.Console.ReadLine());
                }
                // 11.2 catch the conversion error
                catch
                {
                    // 11.3 notify the user of the error
```

```
                    System.Console.WriteLine("You have entered an
invalid selection");
            }

            // 10.3 check for menu selection 1
            switch (iselection) // 12.1
            {
            case 1: // 12.2
            {
                // 1 add a string variable for your name
                String myName = GetName();
                // 4 add + myName to print out your name
                System.Console.WriteLine("Hello World My Name
is:" + myName);

                break;
            }

            // this will demonstrate the capability of arrays
            case 2:            // 13.3
            {
                ArrayDemo();  // 13.2
                break;        // 13.3
            }
            // check to see if we are done with the program
            // change this to use an integer now
            case 9: // 12.3
            {
                System.Console.WriteLine("Thank you and
goodbye!");

                done = true;
                break;
            }
            default: // 12.4 handle the default case
            {
                System.Console.WriteLine("You have made an
invalid selection.");
                break;
            }
```

```
            } // 12.5

        } // 10.4 end of loop code
    }

// show the basic operation of an array
public static void ArrayDemo()
{
    // 13.2 set up a string for each day of the week
    string[] days = new string[]
        {"Sun","Mon","Tue","Wed","Thr","Fri","Sat"};

    int i;

    // 13.3 print out all of the days
    for(i=0;i<=6;i++)
    {
        System.Console.WriteLine("The day of the week is: " +
days[i]);
    }

}
    public static string GetName()
    {
        // 1 add a string variable for your name
        String myName;
        // 2 add in a print statement to tell you to enter your
name

        System.Console.WriteLine("Please Enter Your Name:");
        // 3 now the readline statement will read in your name
        myName = Console.ReadLine();

        return myName;
    }
}
}
```

Link to Array Documentation

For further information and examples of arrays, I have included a link to the C# documentation on arrays from the C# reference.

https://docs.microsoft.com/en-us/dotnet/csharp/programming-guide/arrays/single-dimensional-arrays

Coding Exercise 2

Exercise 2.1

Write a C# program that stores ten even numbers in an array, and prints them out.

Input	Expected Output
Ten place array of the first ten even numbers.	Counting by 2: 2,4,6,8,10,12,14,16,18,20

Exercise 2.2

Write a C# program that finds the largest number in an array.

Input	Expected Output
20,30,35,55,41,65,23,20,10,5	The largest number in the array is: 65

Exercise 2.3

Write a C# program, which sums all the numbers in an array.

Input	Expected Output
1,2,3,4,5,6,7,8,9,10	The sum of the array is: 55

14 Code Outlining in Your Editor

Most code editors have many features that help programmers manage the process of writing code. One convenient feature, for managing large programs, is the outlining of your code. By using this feature, you can collapse large areas of code so that you only see the portion of code you are actively working on. The feature is set up to collapse selected areas of code, and then be able to expand them later when needed. One example use of this, is to be able to expand and collapse code sections that are entire methods. By doing this, you can more easily see the flow of larger sections of code. Let's go ahead and look at an example.

Open Our Last Project

Open the project we just created in Chapter 13. Once you have opened the project, you should see the following. I am going to use the example from the Cloud Shell editor, in the Google Cloud Platform.

Notice the highlighted area that is the small collapsible box. This is the toggle that will allow you to collapse selected areas of code. Notice that these controls also have a dashed line, which indicates the matching beginning and ending areas of the outlining.

```
2
3    namespace HelloWorld
4  ⊟ {
5        // Chapter 13
5        class Program
7        {
8            static void Main(string[] args)
9  ⊟        {
0
1                // 9.1 control variable for program
2                Boolean done = false;
```

Collapse All of the Existing Methods

As practice, to begin using this function, collapse all of the methods in the Main class. Don't worry if you make a mistake; you can always undo your change by using Undo, under the Edit Menu, or CTRL-Z to undo your last action. Also, this does not change any of your code; it just limits the visibility of desired sections of the code.

View the Entire Main Class

Now, having collapsed all of the methods in the Main class, the code should look like this in the editor:

```
using System;

namespace HelloWorld
{ ...
    class Program
    {
        static void Main(string[] args)
        { ...
        }

        // show the basic operation of an array
        public static void ArrayDemo()
        { ...
        }
```

You can now expand these methods if you like using the "+" symbol. Using outlining can help you edit more extended portions of code by being able to limit the visibility of the current section of code you are editing.

15 Using Statements

```
using System;
using System.Collections.Generic;
using System.Linq;
using System.Text;
using System.Threading.Tasks;
```

The Using statement in C# makes a namespace available to your program, without having to fully qualify a call to a method. We have had "using" statements at the top of our projects, but have not addressed them up to this point. By including the correct namespace in the "using" statement, at the top of your C# program, you can save yourself some typing if you are going to heavily use methods from that namespace.

For example:

We have been using the statement: System.Console.Writeline("Hello World"), in our project. If we include the following at the top of our program:

```
using System.Console;
```

We can then invoke the method by using the following:

```
WriteLine("Hello World");
```

By correctly including the "using" statements, we can simplify the coding for our projects. To make sure I have sufficiently illustrated the concept let's go through an example.

Start a New Project

Create a new .NET Core project under your cscode directory. Note that the .NET command will create a directory along with the project.

Linux

```
dotnet new console -name ch15
```

Windows

```
dotnet new console -name ch15
```

Create Hello World

Once the program is created and you have loaded it into your editor, you should see the following code. Go ahead and modify the WriteLine statement to print out your name, as in the following code:

```
using System;

namespace ch15
{
    class Program
    {
        static void Main(string[] args)
        {
            Console.WriteLine("Hello World, my name is Eric!");
        }
    }
}
```

Now run the program to make sure it is working correctly.

```
dotnet run
```

You should see the following output:

```
efrick@cloudshell:~/cscode/ch15
(my-second-project-194921)$ dotnet run
Hello World, my name is Eric!
efrick@cloudshell:~/cscode/ch15
(my-second-project-194921)$
```

Modify the Using Statement

Now add the following line to your program to define System.Console, and modify the WriteLine Statement as shown in the following example:

```
using System;
using static System.Console; // 15.1

namespace ch15
{
    class Program
    {
        static void Main(string[] args)
        {
            // 15.2 simplify write statement
            WriteLine("Hello World, my name is Eric!");
        }
    }
}
```

Run The Program

Now rerun the program, to make sure there are no errors. You should have the same output as in the previous run.

Summary

In this section, I have shown you how you can simplify your programs with the "using" statement. It can be beneficial once you start coding more extensive programs.

Documentation on Using Statements from Microsoft

I have included a link where you can find more information on using statements from the Microsoft C# documentation.

https://docs.microsoft.com/en-us/dotnet/csharp/language-reference/keywords/using-directive

16 File Output

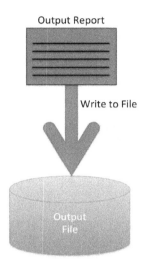

Output Report

Write to File

Output File

One of the most common tasks of a computer program is to write an output file. C# has excellent support for all file input and output operations. The sequence of events to write a file is as follows:

1. Open or create a new file for file output. You also need to specify the directory, file name, and the file type. (In this chapter, I illustrate how to write an ASCII text file.)
2. Write all of the data you need to put into the file with write statements.
3. Close the file after all the output is complete.

Copy Your Program

Make a copy of your last program from Chapter 13, and copy to Chapter 16. Please note, we did create a different program in Chapter 15, so use Chapter 13 instead. Use the following commands:

Linux (GCP)

```
cp -r ./ch13 ./ch16
cd ./ch16
```

Windows

```
xcopy ch13 ch16
cd ch16
```

Once you have loaded your program into the editor, you should see the following code:

```
Boolean done = false;
 int iselection = 0;

 while (done == false)
 {
     System.Console.WriteLine("1) Select to enter in your
name");
     System.Console.WriteLine("2) Array demo");
     System.Console.WriteLine("9) Select to exit to program");
     System.Console.WriteLine("Please Enter Your Selection:");
```

Add Menu Selection

First add a new menu selection, as illustrated in the following code:

```
Boolean done = false;
int iselection = 0;

while (done == false)
{
    System.Console.WriteLine("1) Select to enter in your
name");
    System.Console.WriteLine("2) Array demo");
    System.Console.WriteLine("3) File output demo");
    System.Console.WriteLine("9) Select to exit to program");
    System.Console.WriteLine("Please Enter Your Selection:");
```

Next, add code to handle the new selection. Note that the call to the
FileDemo method is not defined yet, since we have not yet added this code.

```
// this will demonstrate the capability of arrays#
case 2:
    {
        ArrayDemo();
        break;
    }

// this will demonstrate how to write an output file in C#
case 3:
    {
        FileDemo();
        break;
    }
```

Add FileDemo Method

Now add a method as a holding place for our file output code. This type of method is called a stub, and is a placeholder for future code. Place this method just before the ArrayDemo method.

```
// demonstrate how to write an output file in C#
public static void FileDemo()
{

}

// show the basic operation of an array
public static void ArrayDemo()
{
...
```

Code for Writing File

Now we can add the code to create an output file in C#, and complete our stub method.

```
// 16.1 demonstrate how to write an output file in C#
public static void FileDemo()
{
    // 16.2 Create a string array that consists of three lines.
    string[] lines = { "First line", "Second line", "Third line" };


    // 16.3 open a new file for the output
    using (System.IO.StreamWriter file =
        new System.IO.StreamWriter(@"output.txt"))
    {
        // 16.4 now write all of the array to the file
        foreach (string line in lines)
        {
            file.WriteLine(line);
        }
    }

}
```

In the previous code, we have added the following:

16.2 We created a string array to write to our output file.

16.3 We now create the file. Note, the program creates the file in the directory where our program is running.

16.4 We write all of the strings in the array to the output file. Since the { and } surround the entire piece of code in a using statement, the program automatically closes the file.

Run the Program

Now run the program, and select the file demo from the menu:

```
dotnet run
```

Now select the file output demo (3) from the menu:

```
efrick@cloudshell:~/cscode/ch16
(my-second-project-194921)$ dotnet run
1) Select to enter your name
2) Array demo
3) File output demo.
9) Select to exit the program
Please Enter Your Selection:
3
1) Select to enter your name
2) Array demo
3) File output demo.
9) Select to exit the program
Please Enter Your Selection:
9
Thank you and goodbye!
```

Verify the Output

After you have run the program, you should have a new file in the directory called output.txt. You should verify that this file is present and has the correct contents. You can do this by using the following commands on the command line.

Linux (GCP)

```
ls -l
```

Windows

```
dir
```

```
efrick@cloudshell:~/cscode/ch16
(my-second-project-194921)$ ls -l
total 24
drwxr-xr-x 3 efrick efrick 4096 Jan 26 13:29 bin
-rw-r--r-- 1 efrick efrick  170 Jan 26 14:21
ch4.csproj
drwxr-xr-x 3 efrick efrick 4096 Jan 26 14:29 obj
-rw-r--r-- 1 efrick efrick   34 Jan 26 14:29
output.txt
-rw-r--r-- 1 efrick efrick 4753 Jan 26 14:35
Program.cs
```

Also, verify the contents of the file:

Linux (GCP)

```
cat output.txt
```

Windows

```
type output.txt
```

You should see the following lines in the file.

```
efrick@cloudshell:~/cscode/ch16
(my-second-project-194921)$ cat output.txt
First line
Second line
Third line
```

If your output matches, you have successfully created your file from C#. You can now save your file and close out your editor for this lesson.

Complete Code

In case you have any trouble here, you can download the code from the online class, which I included the link for in the introduction of this book. The online class contains all of the code from each chapter.

Summary

In this chapter, I have shown you the basic operations for writing an output text file in C#. Writing files is one of the most common tasks in computer programming. Performing this operation in C# is very straight forward. I should note that, there are other variations of how you can write files, but the one I presented here is the basic version.

Documentation on How to Write a File from Microsoft

I have included a link to file output documentation from Microsoft.

https://docs.microsoft.com/en-us/dotnet/csharp/programming-guide/file-system/how-to-write-to-a-text-file

17 File Input

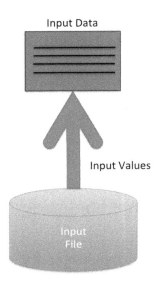

Another common task in computer programming is to read input from a file. In this chapter, you will create an input file,read through the file, and display that input to the screen, one line at a time. This type of input is a common way for programs to read in initialization data, for example.

The typical flow for this task is as follows:

1. Open a file for input, based on the file name and directory.
2. Read data until the end of the file is encountered.
3. Do some processing with the data that the program has read.
4. Close the file when you have completed processing all of the data.

Copy Your Program

Make a copy of your last program from Chapter 16, and move the copy to Chapter 17. Use the following commands:

Linux (GCP)

```
cp -r ./ch16 ./ch17
cd ./ch17
```

Windows

```
xcopy ch17 ch18
cd ch8
```

Once you have loaded the Chapter 17 program into your editor, you should see the following:

```
Boolean done = false;
int iselection = 0;

while (done == false)
{
    System.Console.WriteLine("1) Select to enter in your
name");
    System.Console.WriteLine("2) Array demo");
    System.Console.WriteLine("3) File output demo");
    System.Console.WriteLine("9) Select to exit to program");
    System.Console.WriteLine("Please Enter Your Selection:");
```

Add a Menu Selection

First, add a new menu selection, as illustrated in the following code. This statement is item 4 for the file input demo.

```
Boolean done = false;
int iselection = 0;

while (done == false)
{
    System.Console.WriteLine("1) Select to enter in your
name");
    System.Console.WriteLine("2) Array demo");
    System.Console.WriteLine("3) File output demo");
    System.Console.WriteLine("4) File input demo");
    System.Console.WriteLine("9) Select to exit to program");
    System.Console.WriteLine("Please Enter Your Selection:");
```

Next, add code to handle the new selection. Note that the call to the FileDemo method is not defined, since we have not yet added this code. This code is under the case 4 block.

```
// this will demonstrate how to write an output file in C#
case 3:
    {
        FileDemo();
        break;
    }

// this will demonstrate how to read an input file in C#
case 4:
    {
        FileInputDemo();
        break;
    }
```

Add FileInputDemo Method

Now add a method as a holding place for our file input code. Place this method just before the FileDemo method.

```
// demonstrate how to read an input file in C#
// and display it to the screen
public static void FileInputDemo()
{

}

// demonstrate how to write an output file in C#
public static void FileDemo()
{
    ...
```

Code for Reading a File

Now we can add the code, which reads a text file in C#, and complete the FileInputDemo Method.

```csharp
// demonstrate how to read an input file in C# and
// display it to the screen
public static void FileInputDemo()
{
    int counter = 0;
    string line;

    // 17.1 open the file we created in the previous demo
    System.IO.StreamReader file =
        new System.IO.StreamReader(@"output.txt");

    // 17.2 now loop through every line in the file
    while ((line = file.ReadLine()) != null)
    {
        // 17.3 print the line out
        System.Console.WriteLine(line);
        counter++;
    }

    // 17.4 we are done close out the resource
    file.Close();
    System.Console.WriteLine("There were {0} lines.",
counter);

}
```

In the previous code we have added the following:

17. 1 First, open the file by using the StreamReader file function that is defined in System.IO.

17.2 Next, after opening the file, use a "while loop", which will read one line at a time, until the program reaches the end of the file.

17.3 In the while loop, write out the line that was just read in from the file, to the screen. Then increment the counter variable.

17.4 Finally, once the loop has finished processing all the lines, close the file, and write the number of lines in the file, to the screen.

Run the Program

Now that we have coded our FileInputDemo method, we can run the program by issuing the following command:

```
dotnet run
```

You should see the following output from the program:

```
efrick@cloudshell:~/cscode/ch17 (my-second-project-194921)$
dotnet run
1) Select to enter your name
2) Array demo
3) File output demo.
4) File input demo
9) Select to exit the program
Please Enter -r Your Selection:
4
First line
Second line
Third line
There were 3 lines.
1) Select to enter your name
2) Array demo
3) File output demo.
4) File input demo
9) Select to exit the program
Please Enter -r Your Selection:
9
Thank you and goodbye!
Check the Output
```

If your output matches the above screenshot, you have successfully read in your file using C#, and displayed it on the screen. You can now save your project and close Visual Studio.

Complete Code

In case you have any trouble here, you can download the code from the online class that I have included the link for, in the introduction of this book. The online class contains the code from each chapter, which you can download, if needed.

Summary

In this chapter I have shown you the basic operations of reading an input text file in C#. Reading files is one of the most common tasks in computer programming. Performing this operation in C# is straight forward. I should note, there are other variations of how you can read input files, but the one I have presented here is the basic version.

Documentation on Reading a File from Microsoft

The following link provides further information, from the Microsoft C# documentation, on file input examples:

https://docs.microsoft.com/en-us/dotnet/csharp/programming-guide/file-system/how-to-read-from-a-text-file

18 Strings and Functions

String manipulation is one of the most common operations in a computer program. Strings are used to store all types of common data, such as people's names, email addresses, titles, department names, and the list goes on and on. String functions included in programming languages allow programmers to manipulate strings easily, and to perform various operations on them. Some of these operations include input validation, limiting the length of strings that the user enters into a particular field, and checking that strings follow a required pattern, such as an email address. In this chapter, I demonstrate some of the most commonly used string functions in C#, and how you might use them in a program. You will write a short method that demonstrates some of the most commonly used functions, and you can see how you might be able to use them in your future programs. Also, I have included a link to the reference Microsoft provides, and it's documentation for the complete listing of string functions, which are available.

Information Technology Essentials Volume 2

Copy Your Program

Make a copy of your last program from Chapter 17. Move the copy to Chapter 18. Use the following commands:

Linux (GCP)

```
cp -r ./ch17 ./ch18
cd ./ch18
```

Windows

```
xcopy ch17 ch18
cd ch18
```

```
Boolean done = false;
 int iselection = 0;

 while (done == false)
 {
     System.Console.WriteLine("1) Select to enter in your
name");
     System.Console.WriteLine("2) Array demo");
     System.Console.WriteLine("3) File output demo");
     System.Console.WriteLine("4) File input demo");
     System.Console.WriteLine("9) Select to exit to program");
     System.Console.WriteLine("Please Enter Your Selection:");
```

Add Menu Selection

First, add a new menu selection as illustrated in the following code:

```
Boolean done = false;
  int iselection = 0;

  while (done == false)
  {
      System.Console.WriteLine("1) Select to enter in your
name");
      System.Console.WriteLine("2) Array demo");
      System.Console.WriteLine("3) File output demo");
      System.Console.WriteLine("4) File input demo");
      System.Console.WriteLine("5) String function demo");
      System.Console.WriteLine("9) Select to exit to program");
      System.Console.WriteLine("Please Enter Your Selection:");
```

Next, add code to handle the new selection.

```
// this will demonstrate how to read an input file in C#
case 4:
    {
        FileInputDemo();
        break;
    }

// this will demonstrate how to use string functions in C#
case 5:
    {
        StringDemo();
        break;
    }
```

Now add a method, as a holding place, for our string demo code. Place this method just before the FileInputDemo method.

```
// demonstrate the most commonly used methods in
// the String class and how they are used
public static void StringDemo()
{

}

// demonstrate how to read an input file in C# and display it
to the screen
public static void FileInputDemo()
{
 . . .
```

Code for StringDemo Method

Now that we have our holding place for the StringDemo method, we can add the detailed code that will demonstrate most of the commonly used string functions in C#. Note that there are other functions available in the String class in C#, but the functions that I have included here, represent some of the most commonly used functions. Please refer to the documentation that I've included, to review the complete list of string functions, and experiment with adding some of your own as well.

```csharp
// demonstrate how to use string functions in C#
public static void StringDemo()
{
    String my_string = "This is my string example";

    // 18.1 how to get the string length
    int ilen;
    ilen = my_string.Length;
    System.Console.WriteLine("This string length is: " + ilen);

    // 18.2 convert a string to uppercase
    String new_string = my_string.ToUpper();
    System.Console.WriteLine("my_string in upper case: " +
new_string);

    // 18.3 convert a string to lowercase
    new_string = my_string.ToLower();
    System.Console.WriteLine("my_string in lower case: " +
new_string);

    // 18.4 check the starting string character
    if (my_string.StartsWith("Th"))
    {
        System.Console.WriteLine("This string starts with TH.");
    } else
    {
        System.Console.Write("This string does not start with TH.");
    }

    // 18.5 replace a character in a string
    new_string = my_string.Replace("s", "x");
    System.Console.WriteLine("new_string after replacement:" +
new_string);

        // 18.6 get a substring from the original string
    new_string = my_string.Substring(0, 10);
    System.Console.WriteLine("The first 10 characters of my_string:
"+new_string);

}
```

This code demonstrates the following:

18.1 Get the length of a string.

18.2 Force a string to be all uppercase characters.

18.3 Force a string to be all lowercase characters.

18.4 Check to see if a string starts with a specified substring.

18.5 Demonstrate how to replace a character in a string.

18.6 This line demonstrates the Substring function.

Run the Program

Now that we have coded our StringDemo method, we can run the program by issuing the following command:

```
dotnet run
```

Run the program and select the file string demo from the menu, and you should see the following output:

```
efrick@cloudshell:~/cscode/ch18 (my-second-project-194921)$
dotnet run
1) Select to enter your name
2) Array demo
3) File output demo.
4) File input demo
5) String function demo
9) Select to exit the program
Please Enter -r Your Selection:
5
This string length is: 25
my_string in upper case: THIS IS MY STRING EXAMPLE
my_string in lower case: this is my string example
This string starts with TH.
new_string after replacement:Thix ix my xtring example
The first 10 characters of my_string: This is my
1) Select to enter your name
2) Array demo
3) File output demo.
4) File input demo
5) String function demo
9) Select to exit the program
Please Enter -r Your Selection:
9
Thank you and goodbye!
```

Check the Output

If your output matches the previous example, then you have successfully
written the StringDemo method. You can see, from this example, how easy
it is to manipulate strings from within C#, and how useful they can be
within your future programs.

Summary

In this chapter, I have demonstrated some of the most common operations of string processing within C#. The operations I have shown here are only a minor subset of all the operations that are available, but they show how easily you can integrate them into your program.

Documentation on String Functions from Microsoft

The following link provides further information from the Microsoft C# documentation on String functions:

https://msdn.microsoft.com/en-us/library/system.string_methods(v=vs.110).aspx

Coding Exercise 3

Exercise 3.1

Write a C# program that reads in a string and prints out the length of that string.

Input	Expected Output
This is my input string	The length of the string is 23 characters.

Exercise 3.2

Write a C# program that prints out the first and last character of the input.

Input	Expected Output
This is my input string	The first character is: T The last character is: g

Exercise 3.3

Write a C# program, which converts all the characters of the input to uppercase.

Input	Expected Output
This is my input string	My input in upper case: THIS IS MY INPUT STRING

19 Dates and Functions

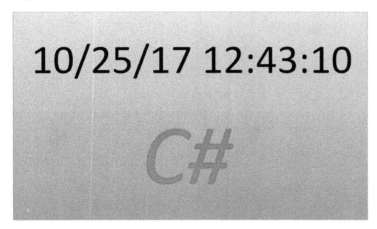

Manipulating dates and calculations around dates and times are one of the essential operations in business-related programs. Dates are used to calculate interest on payments; deadlines for paying bills; and recording demographic information on people, such as birth dates, to name a few examples. Older programming languages did not have built in facilities for manipulating dates. This lack of language support caused many problems in the past, such as the massive rewriting of code that was required for the year 1999 to the year 2000 cutover. Due to the way that programs handled dates in older style programs, much of the program logic in these programs had to be patched, or re-written, to accommodate the changeover of the century. Thankfully, modern programming languages, such as C#, have built-in methods and functions for manipulating dates. In this chapter, we will write a simple function, which illustrates some of the most commonly used functions and how to implement them in C#.

Copy Your Program

Make a copy of your last program from Chapter 18, and move the copy to Chapter 19. Use the following commands:

Linux (GCP)

```
cp -r ./ch18 ./ch19
cd ./ch19
```

Windows

```
xcopy ch18 ch19
cd ch19
```

Once you have loaded the Chapter 19 program into your editor, you should see the following:

```
Boolean done = false;
 int iselection = 0;

 while (done == false)
 {
     System.Console.WriteLine("1) Select to enter in your
name");
     System.Console.WriteLine("2) Array demo");
     System.Console.WriteLine("3) File output demo");
     System.Console.WriteLine("4) File input demo");
     System.Console.WriteLine("5) String function demo");
     System.Console.WriteLine("9) Select to exit to program");
     System.Console.WriteLine("Please Enter Your Selection:");
 }
```

Add Menu Selection

First, add a new menu selection, as illustrated in the following code:

```
Boolean done = false;
 int iselection = 0;

 while (done == false)
 {
     System.Console.WriteLine("1) Select to enter in your
name");
     System.Console.WriteLine("2) Array demo");
     System.Console.WriteLine("3) File output demo");
     System.Console.WriteLine("4) File input demo");
     System.Console.WriteLine("5) String function demo");
     System.Console.WriteLine("6) Date function demo");
     System.Console.WriteLine("9) Select to exit to program");
     System.Console.WriteLine("Please Enter Your Selection:");
```

Next, add code to handle the new selection. You will be adding code for the case 6 selection.

```
// this will demonstrate how to use string functions in C#
case 5:
{
    StringDemo();
    break;
}

// this will demonstrate how to use date functions in C#
case 6:
{
    DateDemo();
    break;
}
```

Add DateDemo Method

Now, add a method as a holding place for our date demo code. Place this method just before the StringDemo method:

```
// demonstrate the most commonly used methods in the
// Date class and how they are used
public static void DateDemo()
{

}

// demonstrate how to read an input file in C# and display it to the
screen
public static void StringDemo()
{
 ...
```

Code for DateDemo Method

Now, add the following code for the DateDemo method:

```
// demonstrate how to use date functions in C#
 public static void DateDemo()
 {
     // 19.1 create a new date October 23, 2017
     DateTime my_date = new DateTime(2017, 10, 23);

     // 19.2 print out the date time in a custom format
     System.Console.WriteLine("The date is: "+ my_date.ToString("MMMM dd,
yyyy"));

     // 19.3 return just the month component
     System.Console.WriteLine("my_date month: " + my_date.Month);

     // 19.4 return just the day component
     System.Console.WriteLine("my_date day of the month: " + my_date.Day);

     // 19.5 get the day of the week
     System.Console.WriteLine("my_date day of the week: " +
my_date.DayOfWeek);

     // 19.6 get the day of the week
     System.Console.WriteLine("my_date day of the year: " +
my_date.DayOfYear);

     // 19.7 create a new date
     DateTime new_date = new DateTime(2017, 10, 28);

     // 19.8 get the number of days between these dates
     int diff;
     diff = (new_date - my_date).Days;
     System.Console.WriteLine("The number of days between: " + diff);

 }
```

Run the Program

Now, run the program with the following command:

```
dotnet run
```

Check the Output

If your output matches the output displayed below, then you have successfully written the DateDemo method.

```
efrick@cloudshell:~/cscode/ch19 (my-second-project-194921)$ dotnet run
1) Select to enter your name
2) Array demo
3) File output demo.
4) File input demo
5) String function demo
6) Date function demo
9) Select to exit the program
Please Enter -r Your Selection:
6
The date is: October 23, 2017
my_date month: 10
my_date day of the month: 23
my_date day of the week: Monday
my_date day of the year: 296
The number of days between: 5
1) Select to enter your name
2) Array demo
3) File output demo.
4) File input demo
5) String function demo
6) Date function demo
9) Select to exit the program
Please Enter -r Your Selection:
9
Thank you and goodbye!
```

Summary

You can see from this example how easy it is to manipulate dates from within C#, and how useful they can be to use in your future programs.

Documentation on Date Functions from Microsoft

The following link provides further information from the Microsoft C# documentation on Date functions:

https://msdn.microsoft.com/en-us/library/system.datetime(v=vs.110).aspx

Coding Exercise 4

Exercise 4.1

Write a C# program that prints out today's date and time in the long format.

Input	Expected Output
Today's current date and time.	Today's date is: 6/18/2019 11:49:01

Exercise 4.2

Write a C# program that prints out the day of the week for the current date and time.

Input	Expected Output
Today's current date and time.	The day of the week is Wednesday.

Exercise 4.3

Write a C# program that prints out the current year, month, and day.

Input	Expected Output
Today's current date and time.	Today the year is: 2019 Today the month is: 6 Today the day is: 19

20 Object-Oriented Programming

Person Class

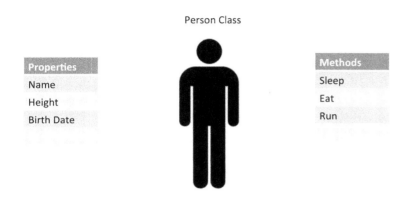

Properties
Name
Height
Birth Date

Methods
Sleep
Eat
Run

Objects in C# allow programmers to more easily model software components after real-world things. In the example above, we have a Person class, which defines the attributes (properties) and actions (Methods) of a Person. By designing your software in this way, it allows your program to more naturally interact with real-world objects. It also makes your programs much easier to understand for yourself, other programmers, and the business subject matter experts, which might be working on your projects.

One key point to understand in object-oriented programming is the difference between a class and an object. A class is the code that represents the entity you are modeling. An object is an instance of a class. In other words, the class is the data type, and the object is the variable.

A source code file can contain multiple classes in one file, but generally, they are organized as one class per file. This makes things easier to find in a large project. Typically, a class might be defined by the following:

```
public class Customer
{
    //Fields, properties, methods and events go
here...
}
```

The following statement can then declare the object of that class:

```
Customer object1 = new Customer();
```

To further illustrate how to define and implement a class fully, as well as an object, we will add a new section to our demo program, By defining and implementing a simple example of object-oriented programming. Let's go ahead and get started.

Copy Your Program

Make a copy of your last program from Chapter 19. Move the copy to Chapter 20. Use the following commands:

Linux (GCP)

```
cp -r ./ch19 ./ch20
cd ./ch20
```

Windows

```
xcopy ch19 ch20
cd ch20
```

Once you have loaded the Chapter 20 program into your editor, you should see the following:

```
Boolean done = false;
int iselection = 0;

while (done == false)
{
     System.Console.WriteLine("1) Select to enter in your
name");
     System.Console.WriteLine("2) Array demo");
     System.Console.WriteLine("3) File output demo");
     System.Console.WriteLine("4) File input demo");
     System.Console.WriteLine("5) String function demo");
     System.Console.WriteLine("6) Date function demo");
     System.Console.WriteLine("9) Select to exit to program");
     System.Console.WriteLine("Please Enter Your Selection:");
```

Add Menu Selection

First, Add a new menu selection, as illustrated, in the following code:

```
Boolean done = false;
int iselection = 0;

while (done == false)
{
    System.Console.WriteLine("1) Select to enter in your
name");
    System.Console.WriteLine("2) Array demo");
    System.Console.WriteLine("3) File output demo");
    System.Console.WriteLine("4) File input demo");
    System.Console.WriteLine("5) String function demo");
    System.Console.WriteLine("6) Date function demo");
    System.Console.WriteLine("7) Object demo");
    System.Console.WriteLine("9) Select to exit to program");
    System.Console.WriteLine("Please Enter Your Selection:");
```

Next, add code to handle the new selection. (Note that we have not yet added the code for our ObjectDemo method.)

```
// this will show how to construct and use a basic object in C#

case 7:
{
    ObjectDemo();
    break;
}
```

Add ObjectDemo Method

Now, add a method as a holding place for our object demo code. Place this method just before the DateDemo method by inserting this code there.

```
// this will show how to construct and use a basic
object in C#
public static void ObjectDemo()
{

}
```

Create Our Person Class

For the first time in this book, we will be adding a new file (and class) to our project. You will need to create a new file in the ch20 directory, with your editor, called Person.cs. Enter the following code into your empty file:

```
using System;

namespace HelloWorld
{
    class Person
    {
    }
}
```

Code Our Person Class

Now that we have created a shell for our Person class, we will add the following code to add some attributes and operations for our class. Enter the following code into our new class.

```
using System;

namespace HelloWorld
{
    // demonstration class to model the basic attributes of a person
    class Person
    {
        // define all person attributes here:
        private int id;
        private String first_name;
        private String last_name;
        private DateTime birth_date;
        private String email_address;
        private String phone_number;

        // this is a basic constructor
        public Person(int new_id,
            String new_first_name,
            String new_last_name,
            DateTime new_birth_date,
            String new_email_address,
            String new_phone_number)
        {
            // set the object attribute to the input parameter values
            this.id = new_id;
            this.first_name = new_first_name;
            this.last_name = new_last_name;
            this.birth_date = new_birth_date;
            this.email_address = new_email_address;
            this.phone_number = new_phone_number;
```

```
        }

        // now create a custom ToString method
        public override string ToString()
        {
            String my_string;

            // create our custom string note the \n is
            // used to create a new line on the screen
            my_string = "Person ID: " + this.id
                + "\nFirst Name: " + this.first_name
                + "\nLast Name: " + this.last_name
                + "\nEmail address: " + this.email_address
                + "\nPhone: " + this.phone_number;

            return my_string;
        }

    }
}
```

A lot is going on in this code. First, note that the entire class is defined by the beginning bracket following the class definition, and is closed by the last bracket in the file. Next, following the class definition, are the definitions of all the attributes for a person. Each of these attributes is defined as a private variable. By defining these as private, it will mean that the client program will not be able to manipulate these attributes directly. Although this may not be very clear to you at this point, it is a best practice to mark attributes as private, so that the class hides the details from the calling program. In this way, you can hide all of the details of how your Person class is implemented from the calling program, so that if you need to make changes to the class, it will minimize the impact on the college program in the future. We will look at this in more detail in our next example, in the next chapter.

The next section of the class is a method called a constructor. This "constructor" is used to initialize the values of an object. I have used the terms "object" and "class" in this chapter, which can be a confusing subject for many beginner programmers. A "class" is merely the code that is used to define an object. An "object" is an instantiation of a class. You can also think of an object as a variable that is defined to be a type of a particular class. We will see the instantiation of an object, or the creation of an object, in the next section of code, which we will add to our demo program.

The last section of the class is an override of an object default ToString method. By using this, we can customize the default behavior of the ToString method, and custom build the string precisely the way we would like the method to return the string. In this example, I have added labels for each of the fields, as well as returning the value of each of them, so the calling program has access to all of the data in this class. Please note that the \n, which I have added in front of all of the field labels, is to force a new line character when the program writes the string to the screen. This technique is a method that Windows programs commonly use.

Code for ObjectDemo Method

The following is the code for our ObjectDemo method. This code creates a new person object, and then calls the constructor, with all the required variables, to create our new person object. The code then prints out the person object, using the ToString method provided by our class. Note that there are other methods we can add to the Person class, to give more functionality, but I have kept this example as simple as possible to illustrate the primary approach used to create a simple class. The following is the listing of the code.

```
// demonstrate how to create a basic object in C#
public static void ObjectDemo()
{
    Person my_person = new Person(1,
        "Eric", "Frick", new DateTime(2017, 1, 1),
        "sales@destinlearning.com",
        "111-222-3333");

    // now print out our newly created object
    System.Console.WriteLine("My new person object: "
    + my_person.ToString());

}
```

Run the Program

Now that we have coded our file, ObjectDemo method, we can run the program with the following command:

```
dotnet run
```

Now, run the program. Select the file object demo (7) from the menu, and you should see the following output:

```
efrick@cloudshell:~/cscode/ch20 (my-second-project-194921)$ dotnet
run
1) Select to enter your name
2) Array demo
3) File output demo.
4) File input demo
5) String function demo
6) Date function demo
7) Object demo
9) Select to exit the program
Please Enter -r Your Selection:
7
My new person object: Person ID: 1
First Name: Eric
Last Name: Frick
Email address: sales@destinlearning.com
Phone: 111-222-3333
1) Select to enter your name
2) Array demo
3) File output demo.
4) File input demo
5) String function demo
6) Date function demo
7) Object demo
9) Select to exit the program
Please Enter -r Your Selection:
9
Thank you and goodbye!
```

If your output matches the above figure, then you have successfully created
your first object in C#. Although this is a simple example, you can quickly
see how this program can easily be extended to add additional
functionality.

Documentation on Classes from Microsoft

The following link provides further information from the Microsoft documentation on classes and how to create them in C#:

https://docs.microsoft.com/en-us/dotnet/csharp/programming-guide/classes-and-structs/classes

21 Inheritance

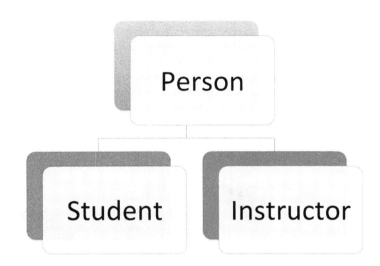

One of the basic principles of object-oriented programming is the concept of inheritance. Inheritance allows you to define a base class, which has a common set of functions and attributes that its children will inherit. The children of the base class can then add extended attributes and functions. In this way, the child classes can share common functionality, but behave slightly different, due to their extended attributes.

In the example above, the base class is a person with two child classes--a Student and an Instructor. In the base class, Person, we have defined a first name, last name, and address. In the child classes, we can add things that are unique to the child classes. For example, in the student class, we could add a grade point average, and in the Instructor class, we could add a pay rate. We could also add many other things that uniquely define Students and Instructors. With these two attributes though, you should get the idea that the Student and the Instructor can share common code and functionality from the Person class, while the inherited classes provide unique attributes for their classes. To further illustrate the point, let's go ahead and define the Instructor class, and add it to our sample program.

Copy Your Program

Make a copy of your last program, from Chapter 19, and move the copy to Chapter 20. Use the following commands:

Linux (GCP)

```
cp -r ./ch20 ./ch21
cd ./ch21
```

Windows

```
xcopy ch20 ch21
cd ch21
```

Once you have loaded the Chapter 21 program, into your editor, you should see the following:

```
Boolean done = false;
 int iselection = 0;

 while (done == false)
 {
     System.Console.WriteLine("1) Select to enter in your
name");
     System.Console.WriteLine("2) Array demo");
     System.Console.WriteLine("3) File output demo");
     System.Console.WriteLine("4) File input demo");
     System.Console.WriteLine("5) String function demo");
     System.Console.WriteLine("6) Date function demo");
     System.Console.WriteLine("7) Object demo");
     System.Console.WriteLine("9) Select to exit to program");
     System.Console.WriteLine("Please Enter Your Selection:");
```

Add Menu Selection

First, add a new menu selection, as illustrated in the following code:

```
Boolean done = false;
int iselection = 0;

while (done == false)
{
    System.Console.WriteLine("1) Select to enter in your name");
    System.Console.WriteLine("2) Array demo");
    System.Console.WriteLine("3) File output demo");
    System.Console.WriteLine("4) File input demo");
    System.Console.WriteLine("5) String function demo");
    System.Console.WriteLine("6) Date function demo");
    System.Console.WriteLine("7) Object demo");
    System.Console.WriteLine("8) Inheritance demo");
    System.Console.WriteLine("9) Select to exit to program");
    System.Console.WriteLine("Please Enter Your Selection:");
```

Next, add code to handle the new selection. Note that the call to the InheritanceDemo method is not yet defined, since we have not yet added this code.

```csharp
// this will how to construct and use a basic object in C#
case 7:
{
    ObjectDemo();
    break;
}

// this will demonstrate how to utilize inheritance within C#
case 8:
{
    InheritanceDemo();
    break;
}
```

Add InheritanceDemo Method

Now, add a method as a holding place for our inheritance demo code. Place this method just before the ObjectDemo method.

```csharp
// this will demonstrate how to utilize inheritance within C#
public static void InheritanceDemo()
{

}

// this will how to construct and use a basic object in C#
public static void ObjectDemo()
{
    ...
```

Create Our Instructor Class

For our examples of inheritance, we will create an Instructor class that will inherit its behavior from our base class, which is the Person class we created in the last chapter. As with creating our Person class, we will need to create a new class and file in our project, calling it Instructor.cs. Create a new file with your editor, and put in the following code:

```csharp
using System;

namespace HelloWorld
{
    class Instructor
    {
    }
}
```

Now that we have a basic shell for our class, we will add the following modifications to make this our inherited Instructor class:

```csharp
using System;

namespace HelloWorld
{
    public class Instructor : Person
    {
        private string title;
        private double hourly_rate;

        // this is a basic constructor
        public Instructor(int new_id,
            String new_first_name,
            String new_last_name,
            DateTime new_birth_date,
            String new_email_address,
            String new_phone_number,
            String new_title,
            double new_hourly_rate)
            : base(new_id,
                new_first_name,
                new_last_name,
                new_birth_date,
                new_email_address,
                new_phone_number)
        {
            // set the object attributes to the input parameter values
            this.title = new_title;
            this.hourly_rate = new_hourly_rate;

        }
    }
}
```

Note that in this class, we have added in two new attributes for our instructor object. We have added a title, which is a string, and an hourly rate – defined as a double. This will allow the code to associate an hourly rate with an instructor. Please note that in the constructor, for the instructor, we need to add the :base portion of the method call. This is done to invoke the constructor of the parent object, which, in this case, is the Person class. You will also note that the only attributes we actually have to set, are the new fields that we added for the title and the hourly rate. By using inheritance, we can easily extend the functionality of base classes, and layer in new functionality for derived objects. By using this technique, you can easily model a complex hierarchy of objects, and support them in your C# programs.

Code for InheritanceDemo Method

The following is the code for our ObjectDemo method. This code creates a new Instructor object, and then calls the constructor, with all the required variables, to create our new instructor object–including our new fields for title and hourly rate. The code then prints out the Instructor object using the ToString method, provided by our base class. Notice, by using this coding technique, we are merely extending the functionality of the base object.

```
// this will demonstrate how to utilize inheritance within C#
public static void InheritanceDemo()
{
    Instructor my_person = new Instructor(1,
        "Eric", "Frick", new DateTime(2017, 1, 1),
        "sales@destinlearning.com",
        "111-222-3333",
        "Computer Science Instructor",
        29.99);

    // now print out our newly created object
    System.Console.WriteLine("My new person object: "
        + my_person.ToString());
}
```

Run the Program

Now that we have coded our file InheritanceDemo method, we can run the program by issuing the following command:

```
donet run
```

Now, run the program and select the file object demo (7) from the menu, and you should see the following output:

```
efrick@cloudshell:~/cscode/ch21 (my-second-project-194921)$ dotnet run
1) Select to enter your name
2) Array demo
3) File output demo.
4) File input demo
5) String function demo
6) Date function demo
7) Object demo
8) Inheritance demo
9) Select to exit the program
Please Enter Your Selection:
8
My new person object: Person ID: 1
First Name: Eric
Last Name: Frick
Email address: sales@destinlearning.com
Phone: 111-222-3333
1) Select to enter your name
2) Array demo
3) File output demo.
4) File input demo
5) String function demo
6) Date function demo
7) Object demo
8) Inheritance demo
9) Select to exit the program
Please Enter Your Selection:
9
Thank you and goodbye!
```

If your output matches the above screenshot, then you have successfully implemented your first inherited object within C#. Again, this is a simple example of inheritance, but you can see how you can easily implement a hierarchy of objects quickly and efficiently, to model real-world classes to use within your programs.

Documentation on Inheritance from Microsoft

The following link provides further information from the Microsoft documentation on inherited classes and how to create them in C#:

https://docs.microsoft.com/en-us/dotnet/csharp/tutorials/inheritance

22 A Complete Example

// complete example

Now that we have completed our demo program, we have the basic building blocks necessary to write a complete application. For our comprehensive program, we will write a simple Rolodex program, to track peoples names and basic contact information. We will, again, write this program as a console program so we can concentrate on the basic C# language constructs and necessary program construction details. We will use most of the previous techniques we have learned so far, to build this program.

Our program will have the following requirements:

- We need to track the first and last name of each person in our Rolodex.
- We also need to track their email address and phone number.
- We need to be able to track a maximum of 100 people in our Rolodex.
- We need to be able to save all of our entries in a file on the computer.
- Our program needs to be able to add, edit, and delete people from our Rolodex.
- Our program needs to be able to display all of the entries on the screen
- Each person in the system must have a unique id assigned to them.

Although these requirements may seem simple, it is essential to pay attention to each one of these to make sure we include them in our final product. Many software development projects fail because they do a poor job of defining the requirements and tracking them throughout the project.

The steps I will show you in building this project are a typical way a modern software development team would tackle this problem. Many teams, today, use an agile methodology, where the team breaks down the project into short steps, or delivery cycles. An example of this, is the Scrum methodology, where the project is broken down into two-week sprints. Don't worry, the steps below are much more straightforward and will not take two weeks, but I want to get you used to breaking down a software project into small, bite-sized pieces, which you can implement, a little bit at a time.

The steps I present below are not the only way to code this, but it represents a typical way of how to implement a system using a top-down approach. The top-down approach will code the highest level functions first, such as the menu, and then implement the next level pieces, one at a time, until all of the required features have been developed . Using this approach is an excellent way to produce demos for customers. You can show them, at any step along the way, what their software looks like so far—making sure there are no surprises. With that in mind, let's go ahead and get started.

Create a New Project

Since we are breaking away from our step-by-step demo program, which we were working with earlier, we will now create a brand-new project, for our complete example. First, create a new directory for our project and navigate to that directory:

Linux (GCP)

```
mkdir ch22
cd ./ch22
```

Windows

```
mkdir ch22
cd ch22
```

After you have created a new directory, create a new .NET Core project:

```
dotnet new console --name Rolodex
```

Once our .NET has finished creating our new project, we will see the code as displayed in the following listing:

```
using System;

namespace Rolodex
{
    class Program
    {
        static void Main(string[] args)
        {
        }
    }
}
```

Create the Main Menu

The first step in creating our project, is to write our main menu for the console application that allows the user to select which operation they would like to perform. This code should be very familiar and is identical, in structure, to the code we developed for the menu in our demo program. If you want to save time, you can copy and paste the code from our previous program, and modify it. Your menu code should look like the following listing:

```
using System;

namespace Rolodex
{
    class Program
    {
        static void Main(string[] args)
        {

            Boolean done = false;
            int iselection = 0;

            while (done == false)
            {
                System.Console.Clear();
                System.Console.WriteLine("1) Add a new name to the Rolodex");
                System.Console.WriteLine("2) Delete a name from the Rolodex");
                System.Console.WriteLine("3) Edit a Rolodex Entry");
                System.Console.WriteLine("4) List all Rolodex Entries");
                System.Console.WriteLine("9) Exit program");
                System.Console.WriteLine("Please Enter Your Selection:");

                // Read in a string and try to convert it to an integer
                //If it can't be converted, capture the error.
                try
                {
                    iselection = Convert.ToInt32(System.Console.ReadLine());
                }
                catch
                {
                 System.Console.WriteLine("You have entered an invalid selection.");
                }
                switch (iselection)
                {

                    // check to see if we are finished with the program
                    // change this to use an integer now
                    case 9:
                        {
                            System.Console.WriteLine("Thank you and goodbye!");
                            done = true;
                            break;
                        }

                    default:
                        {
```

```
                                System.Console.WriteLine("You have made an invalid
selection.");
                           Break;
                    }
                 }
              }
              }
           }
}
```

Note, the above code implements the primary menu and only has code to process the exit selection. This process of coding the essential element first, is called top-down construction. We will implement the most basic function first, then successively add more lower-level details one at a time, until we implement all the requirements of the program. When we run the program, you will only be able to select the exit function. Compile and run this program now, to verify that your basic menu program works.

Create the Stub Methods

Now, we will add stub methods, for all of the menu selection, to give a temporary holding place for all of our program operations. By doing this, we will have a top-level blueprint for our program, and know exactly where to modify our program, as we add functions step by step. Your stub methods should look like the following listing:

```
// add a new person to the Rolodex
public static void add_entry()
{
    System.Console.WriteLine("Add entry selected:");
    System.Console.WriteLine("Press any key to continue:");
    System.Console.ReadKey();
}

// delete a person from the Rolodex
public static void delete_entry()
{
    System.Console.WriteLine("Delete entry selected:");
    System.Console.WriteLine("Press any key to continue:");
    System.Console.ReadKey();
}

// edit an entry in the Rolodex
public static void edit_entry()
{
    System.Console.WriteLine("Edit entry selected:");
    System.Console.WriteLine("Press any key to continue:");
    System.Console.ReadKey();
}

// list all the entries in the Rolodex
public static void list_entry()
{
    System.Console.WriteLine("List entry selected:");
    System.Console.WriteLine("Press any key to continue:");
    System.Console.ReadKey();
}
```

Now, add the case statements after the menu display, in our Main method, to call our stub methods. Your code should look like the following:

```
// process adding a new person
case 1:
{
    add_entry();
    break;
}

// delete a Rolodex entry
case 2:
{
    delete_entry();
    break;
}

// edit an existing entry
case 3:
{
    edit_entry();
    break;
}

// list all entries in the Rolodex
case 4:
{
    list_entry();
    break;
}
```

Create Our Person Class

One of the great things about object-oriented programming, is the ability to reuse existing code, and then being able to extend that code to meet your current needs. In our case, we have already developed a Person class, which should be able to store the base attributes that we are looking for in our Rolodex program. We can borrow the code from our last program, by copying the file to our directory from the previous program.

Linux

```
cp ../ch21/Person.cs .
```

Windows

```
copy \ch21\Person.cs .
```

The only change we need to make, for right now, is to the namespace in the Person.cs file from HelloWorld to Rolodex. This line of code is following the using statements at the top of the file. Change the following line from this:

```
using System;

namespace HelloWorld
{
    // demonstration class to model the basic attributes of a
person
    public class Person
```

To this:

```
using System.Threading.Tasks;

namespace Rolodex
{
    // demonstration class to model the basic attributes of a
person
    public class Person
```

Once you have completed this, use the following command to build (not run) your program, to verify all of your changes have the correct syntax.

```
dotnet build
```

If all succeeds, you will see the Build success message from .NET Core. Now that we have the basic structure of data to hold our Person information, we will soon begin adding the required functions to our program.

Add Main Data Structure

Now that we have our menu structure and Person object in place, we need the primary data structure, to hold our Rolodex entries. We will use an array of 100 people objects to store these. To do this is a two-step process. First, we must declare an array of 100 People objects. Next, we must initialize these objects to be empty people objects. We can initialize this array with the following code, which we are adding at the top of our Main method:

```
static void Main(string[] args)
  {
      // create the array of people objects to
      // store our Rolodex information
      Person[] my_list = new Person[100];

      int i;
      for (i = 0; i < 99; i++)
      {
          my_list[i] = new Person(0,
            String.Empty,
            String.Empty,
            new DateTime(2000, 1, 1),
            String.Empty,
            String.Empty);
      }
```

Build the program, to make sure your new code is correct. You should also run the program to make sure there are no errors.

```
dotnet build
```

Add Method

The first significant function we will add to our system is the capability to add a person to our Rolodex system. We will add a series of print statements, to prompt the user to enter the required information for the Rolodex entry. We will also add matching read statements that read in each data element, as the user enters them from the keyboard, and stores them in variables. These we will then use to add to our array of person objects. Each time the program adds a new person to the Rolodex, we will add it to the next available slot in the array of objects. At this point, we will only be adding our Person objects to the array, which is in local memory, so this means the program will not save the data for the next run. However, in the next step, we will add the "save" method. which will write our array to a data file, in which we can store our Rolodex entries for future use of the program. Listed below is the code for the add_entry method.

```
// add a new person to the Rolodex
public static void add_entry(int rec_to_add)
{
    string fname;
    string lname;
    string email;
    string phone;
    DateTime bdate = new DateTime(2000, 1, 1);

    // 22.2 prompt the operator to enter a new entry into the
Rolodex
    System.Console.WriteLine("Please enter a new entry:");
    System.Console.Write("First Name: ");
    fname = System.Console.ReadLine();
    System.Console.Write("Last Name: ");
    lname = System.Console.ReadLine();
    System.Console.Write("email Address: ");
    email = System.Console.ReadLine();
    System.Console.Write("Phone Number: ");
    phone = System.Console.ReadLine();

    // 22.3 add the new entry to the Person array
    my_list[rec_to_add] = new Person(rec_to_add,
        fname,
        lname,
        bdate,
        email,
        phone);

    // notify the user that the entry has been added
    System.Console.WriteLine("Your new entry has been added.");
    System.Console.WriteLine("Press any key to continue.");
    System.Console.ReadKey();

}
```

The above code features the following:

22.1 The first part of the code declares some local variables to store the Rolodex entry.

22.2 Next, there is a series of write and read statements to read in each field in the record.

22.3 Finally, the new Person object is added to the array, and the user is notified.

You will also have to add the following code, to the Main method, to pass the correct argument to the add_entry method:

```
switch (iselection)
{
    // process adding a new person
    case 1:
    {
        add_entry(next_entry);
        next_entry++;
        break;
    }
}
```

Once you have successfully entered the code for the add_entry method, run the program from the run-without-debugging selection in the run menu. When the program compiles and displays on the screen, select the add selection (1) from the menu and enter the record, as in the following screenshot. Since there is no implementation for the list or the save operations, there is no easy way to test our new add_entry method. Just enter a test record or two to make sure the program is running and not throwing any errors. We will get a better test, in the next section, when we add our list and save methods.

```
1)    Add a new name to the Rolodex
2)    Delete a name from the Rolodex
3)    Edit a Rolodex Entry
4)    List All Rolodex Entries
9)    Exit Program
Please Enter Your Selection
1
Please Enter in a new entry:
First Name: Eric
Last Name: Frick
Email address: efrick@test.com
Phone Number: 614-222-3333
Your new entry has been added.
Press any key to continue.
```

List Method

Now that we can add data to our program, we will implement a simple list_entry method that will display all of the records from the array to the screen. Not only is this a required function of the system, but it will give us an easy way to verify that our add_entry method is functioning correctly, and that the program is correctly populating each element of our person object. Listed below is the C# code for the list_entry method:

```
// list all the entries in the Rolodex
public static void list_entry()
{
    System.Console.WriteLine("First Name,Last Name,Address,Phone
Number,Birthdate");

System.Console.WriteLine("------------------
-----------------------------------------------------------------");

    for (int i=0;i<99;i++)
    {
        string my_person = my_list[i].id +
            "," + my_list[i].first_name +
            "," + my_list[i].last_name +
            "," + my_list[i].phone_number +
            "," + my_list[i].email_address +
            "," + my_list[i].birth_date.ToShortDateString();

        // make sure there is data in the record
        if (my_list[i].id != -1)
        {
            System.Console.WriteLine(my_person);
        }
    }
    System.Console.WriteLine("Press any key to continue:");
    System.Console.ReadKey();
}
```

Now that we have implemented our list_entry method, we can run the program to test it. First, we can enter a new record, like in the following example:

```
1)    Add a new name to the Rolodex
2)    Delete a name from the Rolodex
3)    Edit a Rolodex Entry
4)    List All Rolodex Entries
9)    Exit Program
Please Enter Your Selection
1
Please Enter in a new entry:
First Name: Eric
Last Name: Frick
Email address: efrick@test.com
Phone Number: 614-222-3333
Your new entry has been added.
Press any key to continue.
```

If your output matches the above example, then you have successfully created the list_entry function in our program. You can now move on to the save_all_entries method so the program will retain data from run to run.

Next, select the list_entry function, and you should see the following output:

```
Please Enter Your Selection
4
First Name,Last Name,Address, Phone Number,Birthdate
-------------------------------------------------------------
------
0,Eric,Frick,efrick@test.com,614-111-2222,1/1/2000
```

Save Method

Now that we can enter data into our Rolodex program, and display it on the screen, the next step is to be able to save the data to a file, so the program can load it for future runs. The code that is used to implement the save_all_entries method, and which is listed below, uses a "for loop" to iterate through all of the person entries, in the array, and write them out one by one to a file. Note, there is a check within the loop to see if the list ID is equal to -1, which means the program has deleted the record. In this case, if the application has removed the record, we will not write it out to the file.

```
// save all the entries in the Rolodex
public static void save_all_entries()
{

    // open the file for output
    System.IO.StreamWriter file = new
System.IO.StreamWriter(@"Rolodex.csv");

        for (int i = 0; i < 99; i++)
        {
            string my_person = my_list[i].id +
                "," + my_list[i].first_name +
                "," + my_list[i].last_name +
                "," + my_list[i].phone_number +
                "," + my_list[i].email_address +
                "," + my_list[i].birth_date.ToShortDateString();

            // make sure there is data in the record
            if (my_list[i].id != -1)
            {
                file.WriteLine(my_person);
            }
        }

    file.Close();

    System.Console.WriteLine("Your file has been saved.");
    System.Console.WriteLine("Press any key to continue.");
    System.Console.ReadKey();
}
```

Once you have entered the code and the program compiles, you can test the application by entering a record and then exiting the program. You can then check in the bin directory, to make sure the file has been created and the entry that you created in your test run, is in the file. The following two screenshots illustrate this:

```
1)     Add a new name to the Rolodex
2)     Delete a name from the Rolodex
3)     Edit a Rolodex Entry
4)     List All Rolodex Entries
9)     Exit Program
Please Enter Your Selection
1
Edit entry selected:
Enter an ID to update:
0
Please Enter in a new entry:
First Name: Eric
Last Name: Frick
Email address: efrick@test.com
Phone Number: 614-222-3333
Your new entry has been added.
Press any key to continue.
```

If you navigate to the bin directory, you can open the Rolodex.CSV file in notepad. It should look like the following screenshot:

```
0,Eric Frick,614-111-2222,efrick@test.com,1/1/200
```

If your file looks like the example, you have successfully implemented the save_all_entries method. You can now move on to the load_all_entries method.

Load Method

Now that we have the program data saved in a file, we can read in the data from the file, when we start our program. The following listing is the code for the load_all_entries method. The program first opens a file and reads through the file, line-by-line, until the end of the file. This code is very similar to the file read example we wrote earlier in the book. The main difference is, that after each line is read in, it is parsed into an array using the string.Split function. This function is convenient for breaking apart strings that are delimited with a particular character. (Which in our case is a comma). After we split the string values, they are then assigned to the fields in our person object and loaded into the correct spot in the array. The counter variable tracks the right place in the array.

```
public static void load_all_entries()
{
    // open the file and read in each line
    int counter = 0;
    string line;

    // open the file we created in the previous demo
    System.IO.StreamReader file =
        new System.IO.StreamReader(@"Rolodex.csv");

    // loop through every line in the file
    while ((line = file.ReadLine()) != null)
    {
        string[] values = line.Split(',') ;
        my_list[counter].id = counter;
        my_list[counter].first_name = values[1];
        my_list[counter].last_name = values[2];
        my_list[counter].email_address = values[3];
        my_list[counter].phone_number = values[4];

        // increment the counter for the next entry in the file
        counter++;
    }

    // make sure to set next_entry after all the names have been
loaded
    next_entry = counter;

    file.Close();
    System.Console.WriteLine("There were {0} names loaded",
counter);

}
```

To test this program, run it, and it should now read in the values from the last run. Once the application has loaded, select the list_entry function, and it should display the contents in the file from the previous run. The following screenshot is an example listing of this new feature:

```
1)     Add a new name to the Rolodex
2)     Delete a name from the Rolodex
3)     Edit a Rolodex Entry
4)     List All Rolodex Entries
9)     Exit Program
Please Enter Your Selection
4
First Name,Last Name,Address, Phone Number,Birthdate
-------------------------------------------------------------
------
0,Eric,Frick,efrick@test.com,614-111-2222,1/1/2000
```

With this now working, our program will save entries from one run to the next, and will function as a useful Rolodex program. We can now move on to the delete_entry and edit functions, to complete our program.

Delete Method

To delete a record in our program, we do not actually need to delete anything. Instead, we can merely use one of the variables in the person object to denote, in the array, that the program has deleted the record. We will use the ID field for this. In the code below, we first prompt the operator to enter the ID of the person to delete. Next, we verify that the user enters a valid integer. If it is a valid number, then we use that as the index of the array and mark that person's id with a -1. This indicates that the program has deleted the record. Both the save and display functions have code in place to not display or save a deleted record.

```
// delete a person from the Rolodex
public static void delete_entry()
{
    int iselection = 0;
    System.Console.WriteLine("Delete entry selected:");

    // read in a string and convert it to an integer
    // validate it can be converted
    try
    {
        iselection = Convert.ToInt32(System.Console.ReadLine());
        // mark the desired record with a -1 id
        // so it will be marked as deleted
        my_list[iselection].id = -1;
        System.Console.WriteLine("Your record has been deleted.");
    }
    catch
    {
        System.Console.WriteLine("You have entered an invalid ID.");
    }

    System.Console.WriteLine("Press any key to continue:");
    System.Console.ReadKey();
}
```

Once you have entered the code correctly, for the delete_entry method, you can test it by deleting an existing record, and then running the list_all_entries function to verify the program removed the record. First, list out the current entries in your program:

```
1)    Add a new name to the Rolodex
2)    Delete a name from the Rolodex
3)    Edit a Rolodex Entry
4)    List All Rolodex Entries
9)    Exit Program
Please Enter Your Selection
4
First Name,Last Name,Address, Phone Number,Birthdate
-----------------------------------------------------------------
------
0,Eric,Frick,efrick@test.com,614-111-2222,1/1/2000
1,Test,Person,614-333-4444,test@test.com,1/1/2000
Press any key to continue:
```

Next, select the desired record to delete:

```
1)     Add a new name to the Rolodex
2)     Delete a name from the Rolodex
3)     Edit a Rolodex Entry
4)     List All Rolodex Entries
9)     Exit Program
Please Enter Your Selection
2
Delete entry selected:
1
Your record has been deleted.
Press any key to continue:
```

Finally, verify your selection is no longer in the list.

Notice how we can test each function immediately after we have added new code. Testing is a good habit to establish early on while you are learning to code. This technique will pay big dividends later, when you are working on larger, more complex projects.

Edit Method

One of the ways you can dramatically increase the quality of your programs is to find ways to reuse existing code. Reusing code will make your programs much easier to maintain, over time. To implement the edit_entry function, we can reuse the add_entry function to edit an existing record. Instead of passing the add_entry function the new open slot in the array, we will pass the index of a current record to the add_entry function and overwrite this record. The edit_entry function first prompts the operator to get the correct record id. Then we pass this argument to the add_entry function, to complete the job.

```
// edit an entry in the Rolodex
public static void edit_entry()
{
    System.Console.WriteLine("Edit entry selected:");
    int iselection = 0;
    System.Console.WriteLine("Enter an ID to update:");

    // read in a string and convert it to an integer
    // validate it can be converted
    try
    {
        iselection = Convert.ToInt32(System.Console.ReadLine());
        add_entry(iselection);
        System.Console.WriteLine("Your record has been updated.");
    }
    catch
    {
        System.Console.WriteLine("You have entered an invalid ID.");
    }

    System.Console.WriteLine("Press any key to continue:");
    System.Console.ReadKey();
}
```

To test this function, list out the records in your program, and then select the ID of the record to edit. Then, enter your changes and re-list the records, to make sure the program saved your change correctly. The following screenshots are an example of testing the edit feature.

1) Select the record to edit and make the changes:

```
1)    Add a new name to the Rolodex
2)    Delete a name from the Rolodex
3)    Edit a Rolodex Entry
4)    List All Rolodex Entries
9)    Exit Program
Please Enter Your Selection
3
Edit entry selected:
Enter an ID to update:
0
Please Enter in a new entry:
First Name: Eric1
Last Name: Frick1
Email address: erick1@test.com
Phone Number: 614-222-3333
Your New entry has been added.
Press any key to continue.
```

2) Verify your changes were made correctly:

```
1)     Add a new name to the Rolodex
2)     Delete a name from the Rolodex
3)     Edit a Rolodex Entry
4)     List All Rolodex Entries
9)     Exit Program
Please Enter Your Selection
4
First Name,Last Name, Address,Phone Number,Birthdate
----------------------------------------------------------
0,Eric Frick1,614-222-3333,efrick1@test.com,1/1/2000
Press any key to continue:
```

Summary

In this section, we have written a complete example, which implements many of the concepts that we have learned in the earlier examples in this book. Although this example is not particularly complicated, it represents a complete working program that has a menu structure, a way to add, edit and update data, and a way to save that data for future runs of the program. It also gives you a very high-level example of how simple requirements can be laid out for a software project, and how to write code to meet each of those requirements.

The last major point of the examples in this chapter, is to present a way to test your program, as you are writing new code. You will find that testing will become a significant part of your work, as you get deeper into software development. Software quality is a critical part of all modern software development efforts.

23 Summary

Thank you so much for reading this book. I hope it has given you a good start with programming in the C# language. You have built a very comprehensive sample program, which contains many of the features that you need, to build more complex programs. The sample program is structured to provide you with each of these features in a step-by-step manner, so that you can learn each of these concepts in a short amount of time. The best way to learn to program is to learn a little bit at a time, and to continue to practice, for an extended period, so you improve day by day. In the introduction of this book, I provide a link to the online version of this class, which you can sign up for. It contains videos and additional information. You can also download the code for any of the programs in this book, from that class.

If you have suggestions for improving this book, I would love to hear from you. Please leave a review and/or contact me at sales@destinlearning.com. You can also sign up for my newsletter at http://destinlearning.com, where I publish updates about new material that I have produced. I have also started a page on my website that has some links of interest for Microsoft .NET developers. You can find this page here:

https://www.destinlearning.com/net-developer-links

Thank you again.

24 About the Author

Eric Frick

I have worked in software development and IT operations for 30 years as a Software Developer, Software Development Manager, Software Architect and as an Operations Manager. For the last five years I have taught evening courses on various IT related subjects at several local universities. In 2015 I founded http://destinlearning.com, and have developed a series of online courses and books that can provide practical information to students on various IT and software development topics.

I currently work as a full-time course author at Linux Academy teaching Google Cloud Platform certification courses. Please come and visit us at http://linuxacademy.com

25 More From Destin Learning

YouTube

Thank you so much for your interest in this book. I hope it has given you a good start in this exciting part of Information Technology. You can see more from my YouTube channel where we are continuing to post free videos about software development. If you subscribe to my channel, you will get updates as I post new material weekly:
http://youtube.com/destinlearning

Destin Learning Website

If you would like to read any of my other books, the following is a link to my website. I am planning to add more books over time, and you can view new and upcoming material here:

https://destinlearning.com

Thank you again and good luck with your future with Information Technology!

Appendix A Coding Exercise Solutions

Exercise 1.1

```
using System;

namespace _1._1
{
    class Program
    {
        // exercise 1.1
        static void Main(string[] args)
        {
            int first_number = 0;
            int second_number= 0;
            int temp_number;

            System.Console.WriteLine("Input the first number:");
            first_number = Convert.ToInt32(Console.ReadLine());

            System.Console.WriteLine("Input the second number:");
            second_number = Convert.ToInt32(Console.ReadLine());

            // do the swap
            temp_number = second_number;
            second_number = first_number;
            first_number = temp_number;

            // print out the results
            System.Console.WriteLine("First number:"+first_number);
            System.Console.WriteLine("Second number:"+second_number);

        }
    }
}
```

Exercise 1.2

```
using System;

namespace _1._2
{
    class Program
    {
        // exercise 1.2
        static void Main(string[] args)
        {
            int output = 0;

            output = -1 +4 * 6;
            Console.WriteLine(output);

            output = (35+5) % 7;
            Console.WriteLine(output);

            output = 14 + -4 * 6 / 11;
            Console.WriteLine(output);

            output = 2 + 15 / 6 * 1 - 7 % 2;
            Console.WriteLine(output);

        }
    }
}
```

Exercise 1.3

```
using System;

namespace _1._3
{
    class Program
    {
        // exercise 1.3
        static void Main(string[] args)
        {
            int i;
            for(i=20;i>0;i--)
            {
                if (i!=1) {
                  Console.Write(i+",");
                } else
                {
                  Console.WriteLine(i);
                }
            }

        }
    }
}
```

Exercise 1.4

```
using System;

namespace _1._4
{
    class Program
    {
        // exercise 1.4
        static void Main(string[] args)
        {
            int i;
            int[] count = new int[10];

            for(i=0;i<10;i++)
            {
                Console.WriteLine("Enter number " + (i+1) + ":");
                count[i] = Convert.ToInt32(Console.ReadLine());
            }

            int total = 0;
            for(i=0;i<10;i++)
            {
                total = total + count[i];
            }

            Console.WriteLine("The sum of the numbers is: " + total);

            double average = total / 10.0;
            Console.WriteLine("The average of the numbers is: " + average);


        }
    }
}
```

Exercise 2.1

```
using System;

namespace _2._1
{
    class Program
    {
        static void Main(string[] args)
        {
            int[] count = new int[10] {2,4,6,8,10,12,14,16,18,20};

            System.Console.WriteLine("Counting by 2:");

            for(int i=0;i<9;i++)
            {
                // remember to add a space to separate the output
                System.Console.Write(count[i]+" ");
            }
        }
    }
}
```

Exercise 2.2

```
using System;

namespace _2._1
{
  class Program
  {
    // Exercise 2.2
    static void Main(string[] args)
    {
        int[] count = new int[10] {20,30,35,55,41,65,23,20,10,5};

        System.Console.WriteLine("The largest number in the array is:");

        int largest = 0;

        for(int i=0;i<9;i++)
        {
          if(count[i] > largest)
          {
              largest = count[i];
          }
        }

        // now print out the largest value
        System.Console.WriteLine(largest);

    }
  }
}
```

Exercise 2.3

```
using System;

namespace _2._1
{
  class Program
  {
      // Exercise 2.3
      static void Main(string[] args)
      {
          int[] count = new int[10] {1,2,3,4,5,6,7,8,9,10};

          System.Console.WriteLine("The sum of the array is:");

          int sum = 0;

          for(int i=0;i<=9;i++)
          {
            sum = sum + count[i];
          }

          // now print out the sum of the array
          System.Console.WriteLine(sum);

      }
  }
}
```

Exercise 3.1

```csharp
using System;

namespace _3._1
{
    class Program
    {
        // Exercise 3.1
        static void Main(string[] args)
        {
            String input_string;

            input_string = new String("This is my input string");

            int ilen;
            ilen = input_string.Length;

            Console.WriteLine("The length of the input string is: " + ilen);
        }
    }
}
```

Exercise 3.2

```
using System;

namespace _3._1
{
    class Program
    {
        // Exercise 3.2
        static void Main(string[] args)
        {
            String input_string;

            input_string = new String("This is my input string");

            int ilen = input_string.Length;

            // a string is an array of characters and we can access them directly
            Console.WriteLine("The first character is: " + input_string[0]);
            Console.WriteLine("The last character is: " + input_string[ilen-1]);
        }
    }
}
```

Exercise 3.3

```
using System;

namespace _3._1
{
    class Program
    {
        // Exercise 3.3
        static void Main(string[] args)
        {
            String input_string;

            input_string = new String("This is my input string");

            // use string function to convert to uppercase
            Console.WriteLine("My input in uppercase");
            Console.WriteLine(input_string.ToUpper());

        }
    }
}
```

Exercise 4.1

```
using System;

namespace _4._1
{
    class Program
    {
        // Exercise 4.1
        static void Main(string[] args)
        {
            DateTime my_date = DateTime.Now;

            Console.WriteLine("Today's date is: " + my_date.ToString("MM/dd/yyyy
hh:mm:ss"));
        }
    }
}
```

Exercise 4.2

```
using System;

namespace _4._1
{
    class Program
    {
        // Exercise 4.2
        static void Main(string[] args)
        {
            DateTime my_date = DateTime.Now;

            Console.WriteLine("The Day of the week is: " + my_date.DayOfWeek);
        }
    }
}
```

Exercise 4.3

```
using System;

namespace _4._1
{
    class Program
    {
        // Exercise 4.3
        static void Main(string[] args)
        {
            DateTime my_date = DateTime.Now;

            Console.WriteLine("Today the year is : " + my_date.Year);
            Console.WriteLine("Today the month is : " + my_date.Month);
            Console.WriteLine("Today the day is : " + my_date.Day);
        }
    }
}
```

www.ingramcontent.com/pod-product-compliance
Lightning Source LLC
Chambersburg PA
CBHW060543060326
40690CB00017B/3585